AGRICULTURE ISSUES AND POLICIES

CONSUMER FOOD COSTS

MEASURING THE FOOD DOLLAR

AGRICULTURE ISSUES AND POLICIES

Additional books in this series can be found on Nova's website
under the Series tab.

Additional E-books in this series can be found on Nova's website
under the E-books tab.

ECONOMIC ISSUES, PROBLEMS AND PERSPECTIVES

Additional books in this series can be found on Nova's website
under the Series tab.

Additional E-books in this series can be found on Nova's website
under the E-books tab.

AGRICULTURE ISSUES AND POLICIES

CONSUMER FOOD COSTS

MEASURING THE FOOD DOLLAR

STEFANEE L. MARTIN
EDITOR

Nova Science Publishers, Inc.
New York

Copyright © 2011 by Nova Science Publishers, Inc.

All rights reserved. No part of this book may be reproduced, stored in a retrieval system or transmitted in any form or by any means: electronic, electrostatic, magnetic, tape, mechanical photocopying, recording or otherwise without the written permission of the Publisher.

For permission to use material from this book please contact us:
Telephone 631-231-7269; Fax 631-231-8175
Web Site: http://www.novapublishers.com

NOTICE TO THE READER

The Publisher has taken reasonable care in the preparation of this book, but makes no expressed or implied warranty of any kind and assumes no responsibility for any errors or omissions. No liability is assumed for incidental or consequential damages in connection with or arising out of information contained in this book. The Publisher shall not be liable for any special, consequential, or exemplary damages resulting, in whole or in part, from the readers' use of, or reliance upon, this material. Any parts of this book based on government reports are so indicated and copyright is claimed for those parts to the extent applicable to compilations of such works.

Independent verification should be sought for any data, advice or recommendations contained in this book. In addition, no responsibility is assumed by the publisher for any injury and/or damage to persons or property arising from any methods, products, instructions, ideas or otherwise contained in this publication.

This publication is designed to provide accurate and authoritative information with regard to the subject matter covered herein. It is sold with the clear understanding that the Publisher is not engaged in rendering legal or any other professional services. If legal or any other expert assistance is required, the services of a competent person should be sought. FROM A DECLARATION OF PARTICIPANTS JOINTLY ADOPTED BY A COMMITTEE OF THE AMERICAN BAR ASSOCIATION AND A COMMITTEE OF PUBLISHERS.

Additional color graphics may be available in the e-book version of this book.

Library of Congress Cataloging-in-Publication Data

Consumer food costs : measuring the food dollar / editor: Stefanee L. Martin.
    p. cm.
  Includes index.
  ISBN 978-1-61470-695-3 (hardcover)
  1. Food prices--United States. 2. Food--United States--Costs. 3. Food industry and trade--United States. I. Martin, Stefanee L.
  HD9004.C67 2011
  338.4'366400973--dc23
                      2011024687

Published by Nova Science Publishers, Inc. † *New York*

CONTENTS

Preface		**vii**
Chapter 1	A Revised and Expanded Food Dollar Series: A Better Understanding of Our Food Costs *Patrick Canning*	**1**
Chapter 2	How Retail Beef and Bread Prices Respond to Changes in Ingredient and Input Costs *Edward Roeger and Ephraim Leibtag*	**67**
Chapter 3	How Much Do Fruits and Vegetables Cost? *Hayden Stewart, Jeffrey Hyman, Jean C. Buzby,* *Elizabeth Frazão and Andrea Carlson*	**105**
Index		**141**

PREFACE

This book examines the increases in marketing costs of U.S.-produced food commodities which have outpaced increases in the payments farmers have received for these commodities over the past 40 years. Economic theory provides several market structures that could explain this trend. A persistent increase in the U.S. food marketing bill over an extended period suggests that something more fundamental may also be behind the trend for food marketing costs to rise faster than farmers' proceeds, such as changes in both the structure of the food marketing system and in the socioeconomic characteristics of food consumers.

Chapter 1- For many years, USDA's Economic Research Service (ERS) has analyzed annual spending by U.S. consumers on domestically produced food. ERS has published findings from this analysis in a series known as the marketing bill, which identifi ed the costs of marketing the raw farm commodities contained in a typical dollar's worth of U.S.-produced food and the share of the typical food dollar going to farmers. Measurement problems, the discontinuation of several underlying data sources, and increased interest in evolving supply chain relationships prompted ERS to replace the old marketing bill series with a new expanded data series. This new series, named the food dollar series, provides a more detailed answer to the question, "For what do our food dollars pay?"

Chapter 2- Periodic spikes in the prices of major field crops and related commodities such as those from 1971 to 1974, 1994 to 1996, and 2006 to 2008 have stimulated questions about how these shocks affect wholesale and retail food prices. To what extent do wholesale food prices respond to changes in the underlying costs of inputs? How much of a change in input costs is passed through to retail prices and how long does it take for such cost changes to pass through?

Chapter 3- Federal dietary guidance advises Americans to consume more vegetables and fruits because most Americans do not consume the recommended quantities or variety. Food prices, along with taste, convenience, income, and awareness of the link between diet and health, shape food choices. This research updates previous estimates of vegetable and fruit prices, and estimates the cost of satisfying recommendations for adult vegetable and fruit consumption in the 2010 Dietary Guidelines for Americans.

In: Consumer Food Costs
Editor: Stefanee L. Martin

ISBN: 978-1-61470-695-3
© 2011 Nova Science Publishers, Inc.

Chapter 1

A REVISED AND EXPANDED FOOD DOLLAR SERIES: A BETTER UNDERSTANDING OF OUR FOOD COSTS[*]

Patrick Canning

ABSTRACT

A new ERS food dollar series measures annual expenditures on domestically produced food by individuals living in the United States and provides a detailed answer to the question "For what do our food dollars pay?" This new data product replaces the old marketing bill series, which was discontinued due to measurement problems and limited scope. The new food dollar series is composed of three primary series, shedding light on different aspects of evolving supply chain relationships. The *marketing bill series*, like the old marketing bill series, identifi es the distribution of the food dollar between farm and marketing shares.

The *industry group series* identifi es the distribution of the food dollar among 10 distinct food supply chain industry groups. The *primary factor series* identifi es the distribution of the food dollar in terms of U.S. worker salaries and benefi ts, rents to food industry property owners, taxes, and imports. To provide even more information about modern food

---

[*] This is an edited, reformatted and augmented version of the United States Department of Agriculture, publication, Economic Research Service Report Number 114, dated February 2011.

supply chains, each of the three primary series is further disaggregated by commodity groupings (food/food and beverage), expenditure categories (total, food at home, food away from home), and two dollar denominations (nominal, real). The input-output methodology behind the new food dollar series and comparisons with the old marketing bill series are presented. Several key findings of the new series are highlighted and discussed.

Keywords: food dollar, farm share, marketing bill, industry value added, primary factor value added, input-output analysis, supply chain analysis.

ABOUT THE AUTHORS

Patrick Canning is a senior economist with the U.S. Department of Agriculture's Economic Research Service.

SUMMARY

For many years, USDA's Economic Research Service (ERS) has analyzed annual spending by U.S. consumers on domestically produced food. ERS has published findings from this analysis in a series known as the marketing bill, which identifi ed the costs of marketing the raw farm commodities contained in a typical dollar's worth of U.S.-produced food and the share of the typical food dollar going to farmers. Measurement problems, the discontinuation of several underlying data sources, and increased interest in evolving supply chain relationships prompted ERS to replace the old marketing bill series with a new expanded data series. This new series, named the food dollar series, provides a more detailed answer to the question, "For what do our food dollars pay?"

The New Food Dollar Series

The new food dollar series is composed of three primary series, each of which provides a different way of slicing the same food dollar to provide a variety of perspectives:

- The *marketing bill series*, like the previous series of that name, identifies the distribution of the food dollar between farm and marketing shares.
  - o This series indicates that the costs of marketing farm commodities to U.S. food consumers were an average of 4 cents higher per consumer food dollar than was previously reported between 1993 and 2006. In 2008, the farm share was almost 16 percent.
- The *industry group series* identifi es the value added from 10 distinct food supply chain industry groups to the food dollar (that is, the marginal contribution of each industry group to the final food product).
  - o The farm and agribusiness share in this series differs from the farm share in the current marketing bill series (and the old marketing bill) in that it does not include nonfarm value added. In 2008, 4.2 cents of the 15.8-cent farm share was value added from nonfarm supply chain industry groups, such as energy, transportation, and financial services.
  - o This series indicates that payments from each food dollar going to the energy industry group approached 7 cents in 2008, an increase of 75 percent since 1998. These estimates are higher than those provided by the old marketing bill series, which only measured direct energy use of food processors, retailers, and foodservice establishments.
- The *primary factor series* identifi es the distribution of the food dollar in terms of U.S. worker salaries and benefi ts, rents to food industry property owners, taxes, and imports.
  - o This series indicates that U.S. worker salaries and benefi ts coming from each food dollar steadily declined from 55 cents to 51 cents between 2001 and 2008.
  - o Imported ingredients, both food and nonfood, accounted for a growing share of the food dollar, climbing from less than 5 cents in 1993 to nearly 8 cents in 2008.

2008 Marketing bill series

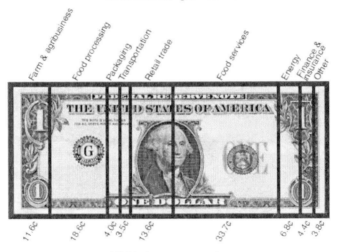

2008 Industry group series

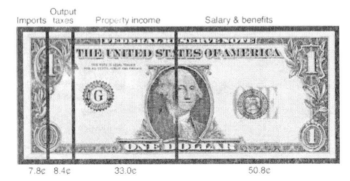

2008 Primary factor series

A Revised and Expanded Food Dollar Series

To provide even more information about food supply chains, each of the three primary series is disaggregated into food-at-home and food-awayfrom-home series and into total food expenditures that do not include soft drinks and alcoholic beverages and total food expenditures that include them. Interestingly, in the food-at-home marketing bill series, the farm share of the food dollar remained around 24 cents from 1993 to 2008, suggesting that increasing expenditures on food services are behind much of the reduction in the farm share in the total marketing bill series.

In total, the new food dollar series includes 36 individual series, created by permutations of the three component series (marketing bill, industry group, primary factor), with the two commodity groupings (food/food and beverage), the three expenditure categories (total, food at home, food away from home), and the two dollar denominations (nominal, real). The series spans the period from 1993 to 2008 and will be updated annually.

How and Why Was the New Food Dollar Series Constructed?

Annual input-output (IO) data for the years 1993 to 2008, published by the Bureau of Labor Statistics; data from the 1997 and 2002 detailed U.S. benchmark IO accounts; and annual IO data for the years 1998 to 2008, published by the Bureau of Economic Analysis (BEA) were compiled and reconciled to produce annual food marketing bill estimates for the period 1993-2008, using conventional JO analysis. Supply chain JO analysis determines where food dollars wind up (as income) by tracing the market value-added measures for 10 supply chain industry groups and three primary production factors (labor, domestic industry assets, and imports). All estimates were reported in both nominal (current price) and real (infl ation-adjusted) dollars.

This new approach to assessing what our food dollars pay for is superior to the former approach in several important ways:

- The quality, timeliness, and completeness of the new source data ensures that a complete accounting of the entire food system is derived from a single consolidated data source.
- A precise approach to measuring and reporting the cost components of the entire food dollar in the new series avoids the potentially confusing divisions of the previous marketing bill series.
- The new food dollar series provides a more complete accounting of the modern global food system.

INTRODUCTION

Increases in marketing costs for U.S.-produced food commodities have outpaced increases in the payments farmers have received for these commodities over most of the past 40 years (Elitzak, 1999, 2004). Economic theory provides several market structures that could explain this trend. For example, a purely competitive market can produce regular fluctuations in the marketing margins of food commodities, driven by population growth and asymmetric supply elasticities for farm commodities and marketing services (Gardner, 1975). Alternatively, when a market segment becomes highly concentrated, collusive pricing strategies among large establishments within this segment can elevate prices, reducing market demand and suppressing prices received by producers who supply commodities to these establishments (Canan and Cotterill, 2006). However, a persistent increase in the U.S. food marketing bill over an extended period suggests that something more fundamental may also be behind the trend for food marketing costs to rise faster than farmers' proceeds, such as changes in both the structure of the food marketing system and in the socioeconomic characteristics of food consumers.

With passage of the Agricultural Marketing Act of 1946, Congress mandated that:

> "The Secretary of Agriculture is directed and authorized to determine costs of marketing agricultural products in their various forms and through the various channels..." [U.S. Code, Title VII, Chapter 38, Section 1622 (b)]

A measure known as the "farm share of the food dollar" was developed by USDA's Economic Research Service (ERS) to meet the mandate for reporting the marketing cost for overall food production. The ERS estimates have come to be called the "food marketing bill."

Beyond the information explicitly called for in the original mandate, the ERS food marketing bill program has historically provided an itemized analysis of annual marketing costs, called "components of the food marketing bill." Figure 1 illustrates the way ERS reported the two data series of the food marketing bill program, the farm share and the marketing bill, in the form of a dollar bill. Each section of the dollar depicts average costs by category for supplying a typical dollar's worth of food to U.S. households. In this format, costs can be expressed interchangeably in terms of cents on a dollar or percentage of total food costs.

A Revised and Expanded Food Dollar Series

ERS food marketing bill estimates have been based largely on a combination of annual data and less frequent Census benchmark statistics that must be adjusted, using conversion factors to conform to the food marketing bill concepts. Over time, the quality and quantity of data for estimating the food marketing bill has diminished, and the method for calculating the marketing bill has become unreliable. A number of authors have discussed food marketing bill estimation issues (Gale, 1967; Harp, 1987; Schluter, Lee, and LeBlanc, 1998; Elitzak, 1999). The aim of the present study is to introduce a systematic method for measuring the marketing bill, using annual data that are being generated on a regular basis.

This report presents the new approach developed by ERS for estimating the food dollar and its component parts.[1] With the new procedures, the complete food dollar, not just the food marketing bill, is divided into total value added for 10 industry groups: farm and agribusiness, food processing, packaging, transportation services, energy, retail trade, food services, finance and insurance, advertising, and legal-accounting-bookkeeping services. Next, the primary factor returns series divides the food dollar by the contributions of three primary production factor groups: domestic hired labor, domestic industry assets, and international imports. Finally, a cross-tabulation table divides the food dollar into the primary factor returns for each industry group. All estimates are reported in nominal (current price) and real (inflation adjusted) dollars. The box "The New Food Dollar Series: A Glossary of Key Terms," presents a list of terms used in the report and their definitions.

Source: USDA, Economic Research Service.

Figure 1. Farm share and marketing bill of the 2006 food dollar computed by the old method.

This new approach to assessing what our food dollars pay for is superior to the former approach in several important ways. First, the quality, timeliness, and completeness of the new source data ensure that a complete accounting of the entire food system is derived from a single consolidated data source. Because the new annual data are largely survey based, year-to-year changes to the supply chain structure, food expenditure patterns, and relative input and output commodity prices are refl ected in these data, unlike in the previous marketing bill series. Second, a precise approach to measuring and reporting the cost components of the entire food dollar in the new series avoids the potentially confusing divisions of the previous marketing bill series. For example, the new food dollar industry group series reports total energy industry costs per dollar of food expenditures, whereas the old marketing bill series only reported the electric and gas utility costs paid by a subset of food marketing establishments, such as processors, retailers, and restaurants. Third, the new food dollar series provides a more complete accounting of the modern global food system. Examples include (1) the explicit measurement of the costs for internationally produced food and nonfood ingredients embodied in all domestically produced food commodities, (2) the explicit food dollar series accounts for food-at-home and food-awayfrom-home expenditures to assess the role of changes to these two distinct food market segments, and (3) the reporting of all food dollar series in both nominal and real (infl ation adjusted) values, to decompose the role of price and volume changes for goods and services embodied in our annual food expenditures. Each of these advantages to the new food dollar series are facilitated by the new source data and input-output analysis methods, and none of these features can be effectively measured using the old data sources and estimation procedures.

A NEW APPROACH TO ESTIMATING AND PRESENTING THE FOOD DOLLAR

Input-output analysis is used to measure the food marketing bill, and supply chain analysis is used to measure the value-added components of the food dollar. To facilitate these new estimation procedures, a precise operational defi nition of food dollar expenditures and the farm share is established.

A Revised and Expanded Food Dollar Series

The New Food Dollar Series: A Glossary of Key Terms

The new and expanded food dollar series uses different source data and estimation methods than the food marketing bill series that it replaces. The changes in methods lead to changes in some terminology. Although many of the key terms below have different meanings in other contexts, only the meaning germane to the context of this report is provided here.

Farm share: Producer value of total annual farm commodity sales that are linked to annual food dollar expenditures, excluding farm commodities that are purchased directly or indirectly by other farm operations. Examples of exclusions are the purchase of hay by a cattle ranch (direct transaction) or purchase by a poultry farm of animal feed containing grains purchased by a feed mill from a feed-grain farm (indirect transaction).

Food dollar: Total annual market value for all purchases of domestically produced foods by persons living in the United States.

Food marketing bill: The market value of all post-farm processes of food dollar supply chain industries, measured as the difference between food dollar expenditures and farm share commodity sales.

Imports: Food and non-food commodities that are imported from international sources and are used by U.S. food supply chain industries producing for the U.S. market.

Industry group: A collection, for accounting purposes, of establishments producing the same or similar output products, and of groups subcontracting to those establishments to support the production of output products.

Industry group value added: The compensation charged by all establishments in an industry group to buyers of their products for the services provided by the industry group's primary factors of production.

Market value: Price paid by a consumer for the purchase of a consumer good.

Nominal value: The dollar value of products and/or services purchased or appraised, based on prevailing prices at the time of purchase or appraisal.

Output taxes: The value of excise, sales, property, and severance taxes (less subsidies), customs duties, and other non-tax Government fees levied on establishments.

Primary factors of production: Assets such as labor, machinery and equipment, physical structures, land and other natural resources, and intellectual property that are employed or operated by an industry group toward fulfi lling the demand for the industry product.

Producer value: Compensation received by producers for the sale of their products.

Real value: The dollar value of products and/or services purchased or appraised, based on prices prevailing during a specifi c time period.

Property income: The pre-tax income or capital gain accruing to owners of non-labor primary factors of production.

Salary and benefits: The pre-tax employee wages plus employer and employee costs for employee benefits.

Supply chain industry: Any industry dedicating resources and/or processes toward fulfi lling the demand for a product.

Today's food marketing system is globally integrated, with many food commodity supply chains having multistage production processes that produce several product and packaging varieties. For example, both domestic and imported wheat is milled in the United States and the flour is widely used by bakeries, who in turn may incorporate fresh fruits and processed ingredients of both domestic and imported origins into their bakery products. In this and many other circumstances, clarity is needed for estimating the costs of marketing U.S. agricultural products, beginning with a concise defi nition of food dollar expenditures. We use the following defi nition:[2]

> Food dollar expenditures are the annual purchases, by people living in the United States, of food products that (1) are produced on a U.S. farm and undergo no off-farm process beyond storage, transport, and basic packaging, or (2) are processed at a domestic food-manufacturing establishment.

As input to these food dollar purchases, farms produce commodities that are either directly consumed as foods or used as ingredients in processed foods.

If we denote this farm production as the *raw food dollar*, then the *farm share* and the *food marketing bill* are defi ned as follows:

> Farm share is measured as the average payment from each food dollar expenditure that farmers receive for their raw food dollar commodities. The food marketing bill is measured as the average value added to the raw food dollar from each consumer food dollar expenditure.

Accounting for Imported Ingredients in Domestically Produced Food

Domestically produced food commodities often rely on imported commodities to facilitate their production, and the food dollar series should include the value added from these imports.

Imported primary (farm fresh) and processed foods purchased by people living in the United States are not included in the proposed defi nition of food dollar expenditures. When used as ingredients by domestic food manufacturers, however, these imports must be treated as a food dollar expenditure. Otherwise, the exclusion of imported wheat used by a U.S. flourmill will erroneously lead to exclusion of domestic fruit filling used by a bakery that

A Revised and Expanded Food Dollar Series

purchased the milled import wheat. Similarly, exclusion of imported fruit filling used by the same bakery will erroneously exclude domestic wheat in other flour purchased by the bakery.

The commingling of imports in domestically produced foods is not limited to food ingredients. Products such as imported petroleum, fertilizers, and transportation equipment are used extensively by domestic establishments producing food. Among the food ingredients that are imported and sold in U.S. markets, those that are commingled with other ingredients and marketed as a product substantially different from the one entering the country are included in the ERS measures of food dollar expenditures, whereas imported food ingredients sold directly in final markets do not enter into the measures.

Measuring Food Dollar Expenditures and the U.S. Farm Share

The first step in the estimation process is to measure average food dollar expenditures and the farm share of those expenditures, using input-output analysis.

Input-output (IO) analysis facilitates the study of interdependencies, both among industries throughout an economy and between industry and final market sales. In the IO framework, an "industry" is a group of establishments that produce similar products, and "fi nal market sales" are all sales of goods or services produced by industries and classifi ed into "commodity groups," other than sales of goods or services for use by a domestic industry for the production of another commodity or service during the current accounting period. A food dollar expenditure is an example of a fi nal market sale. For IO analysis, three subaccounts make up the economic model:

A column[3]Including household purchases of imported foods is necessary for the IO model to trace through the total farm sales linked to food dollar expenditures. These imported food purchases are deducted in a later step.

1) vector y itemizes total fi nal market sales of C distinct commodities, each uniquely produced by one of the C distinct industry groups.
2) A column vector x itemizes total availability of domestic industry output plus imports for each of the C commodity groups.
3) A total requirement matrix, L, also known as the Leontief matrix, is a table with C columns and rows for each industry/commodity group, reporting average annual sales by each industry (such as grain

farming) per dollar of fi nal market demand for each commodity (such as bakery products).

These three subaccounts are related by the simple matrix algebra identity, $L \cdot y = x$, in which multiplication of the fi nal demand vector y by the total requirement matrix L exactly produces the industry output vector x. One convention of IO analysis is the assumption of linearly homogeneous production technologies. A linear technology implies, for example, that if 100 bushels of wheat are required for 9,000 loaves of whole wheat bread sold to U.S. households, then 50 bushels are required for the 4,500 loaves sold to a subset of these households.

Let S_fd denote a column vector that reports the share of each fi nal demand element in y that represents a food dollar expenditure, with the one exception that household purchases of imported foods are included in the share calculation.[3] Next, with subscripts denoting the row and column dimensions of any matrix and with a defi ned as the set of all rows containing agricultural industry/commodity groups, a measure of *import-inclusive gross farm sales* associated with the food dollar expenditures is obtained as follows:

$$x_a^{fd} = L_{a,C} \cdot y_C^{fd}, \quad \text{where } y_C^{fd} = \overline{S\_fd}_C \cdot y_C \tag{1}$$

The *fd* superscript on x_a in equation 1 is to indicate that the x vector is conditional to the multiplication of the final demand vector y by the share vector S_fd. The — symbol above a vector indicates a conversion into a square diagonal matrix. This matrix algebra procedure facilitates row-to-column multiplication of commodity share values in S_fd with the corresponding commodity fi nal demand value in y.

Farm-to-farm payments must be netted out of the gross farm sales measured in equation 1. To explain, note that, in equation 1, gross farm industry sales include interindustry sales by the farm industry, some of which are direct and indirect farm-to-farm sales. For example, a feedlot operation purchases cattle from a ranch (direct) and animal feed from a feed manufacturer that, in turn, purchased grain from a grain farm (indirect). To avoid double counting intraindustry farm sales, one should net out all payments to a farm establishment that are passed on and subsequently go directly or indirectly to another farm establishment:

A Revised and Expanded Food Dollar Series

$$x_a^{net} = x_a^{fd} - \left(\underbrace{A_{a,a}}_{\substack{farm-to-farm \\ direct}} + \underbrace{\hat{A}_{a,a}}_{\substack{farm-to-farm \\ indirect}} \right) \cdot x_a^{fd}$$

(2)

Equation 2 indicates that the import-inclusive net farm sales equal gross farm sales minus the portion of these sales that were purchased by other farm establishments, either directly or through one or more nonfarm industry establishments. In equation 2, $A_{a,a}$ is a matrix describing farm-to-farm direct transactions per dollar of output for each farm commodity, the $A_{a,a}$ matrix with $\wedge$ above it describes total farm-to-farm indirect transactions per dollar of output for each farm commodity,[4] and x_a^{fd} is as defined in equation 1.

To obtain the farm share measure, one must deduct household purchases of imported farm and processed food commodities from the import-inclusive food dollar and the subset of those purchases representing household purchases of imported farm commodities from the net farm sales measured in equation 2. By IO accounting convention, all commodity transactions are divided between domestic and imported sources in proportion to their total availability. If s_m_c denotes the import share of available product for all commodities $c \in C$, the farm share measure that is mandated by Congress is obtained as follows:

$$farm\ share = i_a' \cdot \left[x_a^{net} - \overline{S\_m_a} \cdot y_a^{fd} \right] \div i_C' \cdot \left[\left(i_C - \overline{S\_m_c} \right) \cdot y_C^{fd} \right]$$

(3)

In equation 3, the numerator is a summation of import-exclusive net farm sales and the denominator is a summation of import-exclusive food dollar sales.[5] A detailed mathematical derivation of the expressions used in equations 1 to 3 is provided in a technical appendix to this report.

Annual Food Dollar Estimation Model

The second step is to appropriate data sources to carry out annual estimation of these data series.

The Bureau of Economic Analysis (BEA) publishes a detailed benchmark U.S. input-output table in 5-year intervals, with a 5-year lag between data enumeration and public release of the IO tables (www.bea.gov/industry. Because of their close link to survey-based primary source data and their detailed industry coverage, the BEA benchmark accounts provide the most complete source of information for compiling estimates of equations 1 to 3. The two most recent BEA benchmark IO table releases cover the years 2002 and 1997 and are largely based on the Economic Census data enumerated in those years. After some aggregation of the 1997 and 2002 BEA benchmark accounts to ensure a one-to-one matching of industry groupings, the benchmark tables cover 392 industries.

The Bureau of Labor Statistics (BLS) provides annual input-output tables that are based on the structural matrix of the most recent BEA detailed benchmark IO table (www.bls.gov/emp/ep_data_input_output_matrix.htm). The current BLS annual IO accounts cover calendar year economic flows of the U.S. national economy for 1993 to 2008, and the accounts are reported in both nominal (current-year) and constant (infl ation-adjusted) dollars. Industry output of goods and services is broken into 202 distinct commodity groups, and personal consumption expenditures on food are distinguished by category of purchase.[6] For the present analysis, these categories are broken out into two groups: food and beverages purchased for off-premises consumption (food-at-home), and all other food consumption expenditures (food-away). Examples of food-at-home expenditures include (but are not limited to) food purchased at grocery stores, farmers' markets, or nontraditional food retailers such as convenience stores. Examples of "food-away" expenditures include (but are not limited to) food purchased at restaurants, sports arenas, supplied by employers to employees, and supplied in domestic institutions, for example, school lunches.

The approach in this report is to update the food marketing bill measure obtained from the 2002 detailed benchmark table for the years 2003 to 2008, using the BLS annual IO tables, and to "back-cast" the food marketing bill measure obtained from the 1997 detailed benchmark table for the years 1993 to 1996.

For the years 1998 to 2001, an indexing procedure is employed that captures the relative year-to-year changes in the food marketing bill between 1997 and 2002, as measured by the annual BLS data, while ensuring that the benchmark year estimates are replicated in the index of the BLS series (see Kuchler and Burt, 1990). Documentation of the data development work and the estimation model is provided in the technical appendix to this report.

Documentation of input-output data sources and concepts is available from BEA at www.bea.gov/scb/ pdf/2009/06%20June/0609_ indyaccts_primer _a.pdf .

Historically, alcoholic beverages and soft drinks have not been included in food marketing bill estimates; however, commodities such as wine grapes, hops, and cane sugar are major U.S. crops. Recognizing that certain beverage products such as diet soft drinks use little if any farm commodities, the present analysis nonetheless includes a separate food and beverage dollar series.

Farm Share and Food Marketing Bill Estimates: 1993 to 2008

In the third step, the farm share and food marketing bill series are compiled and analyzed.

Figure 2 presents the 2006 food dollar decomposed into farm share (14.2 cents) and food marketing bill (85.8 cents) components. A comparison with figure 1 highlights two important differences between the revised and the old food marketing bill series. In the 2006 depiction in figure 1, the former series reports a larger farm share value. The result holds true for each year with overlapping measures. The new and the old series use different data sources, which partly explains why the results are different. Beyond this, the step outlined in equation 2 in the new input-output-based estimates, netting out of farm-to-farm direct and indirect transactions, accounts for most of the difference between the old and new farm share estimates over the 1993 to 2006 period where the two series overlap. Farm-to-farm transactions amount to a double-counting of farm sale proceeds of each food dollar, so it is appropriate to adjust for them. The IO accounts provide a systematic means of making these adjustments.

Table 1 reports the estimated farm share and food marketing bill for the U.S. nominal food dollar and the nominal food and beverage dollar, for the period 1993 to 2008, using the new methodology. For the most part, the two series (food versus food and beverage) differ only in levels, with similar year-to-year percentage changes over the study period. For this reason, only the more traditional food dollar series is discussed below.

16   Patrick Canning

Source: USDA, Economic Research Service.

Figure 2. Farm share and marketing share of the food dollar computed by the new method.

In 1993, the farm share of total food dollar expenditures was 18.4 cents, and for the next 4 years it hovered around 18 cents. Beginning in 1998, the farm share of total food dollar expenditures began to decline, reaching 15.3 cents on the food dollar by 2002, a 17-percent drop from the 1998 level. With the exception of 2006 (after the fi rst 2-year decline in farm producer prices in the decade), the 2002 farm share represents a bottoming out of this series over the study period.

The farm share of food-away-from-home expenditures started to decline in 1998, and, by 2002, this share had fallen to 4.8 cents, or half of its 1996 level. For the most part, the 2002 farm share of food-away-from-home expenditures represents a bottoming out, with the exception of a sharp drop in 2006 to 4.1

A Revised and Expanded Food Dollar Series   17

percent, which is the lowest measured level over the study period for both the nominal food dollar and the nominal food and beverage dollar.

Although the farm share of at-home food expenditures does fall off slightly from 1998 to 2002, the series remains above 22 percent and below 24 percent over the study period, with the exception of the fi rst (1993) and final (2007-08) 2 study years. In these 3 years, the farm share of at-home food expenditures was in the low- to mid-24-cent range.

These results present a compelling case that the growing costs and expenditures of eating out are behind the downward trend in the farm share value over the study period. Under this interpretation of the data, the added costs of food preparation and cleanup services in foodservice establishments are driving up the marketing share and driving down the farm share.

To further examine farm share trends over the study period, we compiled the farm share series in constant (infl ation-adjusted) year 2000 prices. Calculations of food dollar expenditures and farm commodity sales associated with these expenditures are compiled as if all transaction prices were fixed at year 2000 levels. Year 2000 estimates will thus be identical to those reported in table 1, whereas estimates for prior and subsequent years will refl ect changes in volumes of purchased food and farm commodities, holding prices constant.[7] Table 2 reports these results.

Between 1993 and 2000, the farm share measure in constant 2000 prices declined at a lower rate than did the nominal measure. This lower rate of decline implies that overall nominal farm commodity prices were falling relative to the prices of goods and services used for marketing these farm commodities to U.S. households. However, starting in 2002, the overall nominal price farm share began to rise relative to the real (infl ation-adjusted) measure, implying that farm commodity prices over this period were rising relative to the prices for food marketing bill goods and services. This rise became more pronounced in 2006 to 2008, coinciding with the upturn in the nominal farm share during this period.

Decomposing the real farm share measure into at-home and food-away expenditures shows that the pre-2000 price trends for farm commodities and for food marketing bill goods and services generally moved together for the at-home expenditures over the 1993 to 2000 period, leading to very similar trends in both the nominal (table 1) and real (table 2) farm share measures over this period. The food marketing bill prices for food-away increased sharply relative to farm commodity prices over this interval. After 2000, farm commodity prices began to gain on food marketing bill prices for at-home food purchases, while food-away marketing bill prices continued to increase relative to the price of farm commodities through 2003. After 2003, farm commodity price increases started outpacing the price of the food marketing bill goods and services for food-away expenditures.

Taken together, the nominal and real farm share measures demonstrate that relative prices are important in shaping the farm share measure but are not the only factor. For example, although farm commodity prices began rising faster than the overall price for food-away marketing bill commodities, the nominal farm share of the food-away dollar trended lower over this interval, except in 2007, when the farm share rose after a sharp decline in 2006. The declining farm share in the face of rising farm commodity prices indicates that the volume of food-away services was increasing, possibly due to the foodservice category's declining relative price. If consumers are eating out more, higher farm commodity prices can coincide with a lower farm share measure due to the added food services purchased per dollar of food expenditures. Further, even if the relative prices for farm commodities and food marketing services remain constant, growth in the share of away-from-home food expenditures would lower the overall farm share measure, since the food-away farm share is substantially lower than the food-at-home farm share.

The prices referred to in the comparisons between nominal and real (infl ation-adjusted) farm share measures represent the implicit per unit costs of a bundle of commodities purchased over the course of a year. Unlike fixed food basket price indexes, this bundle can change over time. For at-home expenditures, a household may change the specifi c products purchased within each commodity group, such as by buying more organic brands or purchasing more food-preparation services like pre-marinated meats. For away-fromhome food expenditures, a household may change food-away destinations from limited-service to full-service establishments. Each of these examples can have food price implications that do not conform to the conventional notion of price infl ation, but are instead caused by year-to-year changes in the product mix purchased within each commodity group. These cost-based price measures should not be confused with conventional commodity price indexes.

Table 1. Marketing bill and farm share of the U.S. nominal food dollar, 1993 to 2008

	Food dollar						Food & beverage dollar*					
	Total		Food at home		Food away		Total		Food at home		Food away	
	Farm share	Market bill	Farm share	Market bill	Farm share	Market bill	Farm share	Market bill	Farm share	Market bill	Farm share	Market bill
					Percent							
1993	18.4	81.6	24.6	75.4	10.5	89.5	16.2	83.8	20.1	79.9	10.1	89.9
1994	17.6	82.4	23.4	76.6	9.6	90.4	15.3	84.7	19.1	80.9	9.1	90.9
1995	18.1	81.9	23.9	76.1	9.7	90.3	15.6	84.4	19.6	80.4	9.0	91.0
1996	17.9	82.1	23.3	76.7	9.6	90.4	15.2	84.8	19.1	80.9	8.5	91.5
1997	17.8	82.2	23.3	76.7	9.4	90.6	15.1	84.9	19.1	80.9	8.5	91.5
1998	17.0	83.0	22.7	77.3	8.2	91.8	14.4	85.6	18.6	81.4	7.3	92.7
1999	16.2	83.8	22.3	77.7	6.9	93.1	13.8	86.2	18.2	81.8	6.1	93.9
2000	15.9	84.1	22.3	77.7	6.2	93.8	13.5	86.5	18.2	81.8	5.5	94.5
2001	15.5	84.5	22.1	77.9	5.5	94.5	13.1	86.9	18.0	82.0	4.9	95.1
2002	15.3	84.7	22.1	77.9	4.8	95.2	12.8	87.2	18.1	81.9	4.1	95.9
2003	15.4	84.6	22.3	77.7	5.1	94.9	12.9	87.1	18.2	81.8	4.5	95.5
2004	15.4	84.6	22.8	77.2	5.0	95.0	13.2	86.8	18.5	81.5	4.6	95.4
2005	15.3	84.7	22.5	77.5	5.0	95.0	13.0	87.0	18.3	81.7	4.5	95.5
2006	14.2	85.8	22.2	77.8	4.1	95.9	12.6	87.4	17.9	82.1	4.1	95.9
2007	15.8	84.2	24.0	76.0	4.8	95.2	13.7	86.3	19.4	80.6	4.6	95.4
2008	15.8	84.2	24.3	75.7	4.7	95.3	14.0	86.0	19.7	80.3	4.6	95.4

*Includes soft drinks and alcoholic beverages.
Source: USDA, Economic Research Service.

Table 2. Marketing bill and farm share of the U.S. real food dollar, 1993 to 2008

| | Food dollar | | | | | | Food & beverage dollar* | | | | | |
| | Total | | Food at home | | Food away | | Total | | Food at home | | Food away | |
	Farm share	Market bill	Farm share	Market bill	Farm share	Market bill	Farm share	Market bill	Farm share	Market bill	Farm share	Market bill
						Percent						
1993	17.1	82.9	22.9	77.1	8.5	91.5	14.7	85.3	18.9	81.1	7.8	92.2
1994	15.3	84.7	20.6	79.4	7.2	92.8	13.1	86.9	16.9	83.1	6.6	93.4
1995	16.2	83.8	21.9	78.1	7.5	92.5	13.9	86.1	18.0	82.0	6.9	93.1
1996	17.0	83.0	22.4	77.6	8.3	91.7	14.2	85.8	18.3	81.7	7.2	92.8
1997	16.3	83.7	21.6	78.4	8.0	92.0	13.8	86.2	17.7	82.3	7.1	92.9
1998	16.1	83.9	21.6	78.4	7.6	92.4	13.5	86.5	17.7	82.3	6.6	93.4
1999	16.1	83.9	22.1	77.9	6.8	93.2	13.6	86.4	18.1	81.9	6.0	94.0
2000	15.9	84.1	22.3	77.7	6.2	93.8	13.5	86.5	18.2	81.8	5.5	94.5
2001	15.8	84.2	22.2	77.8	6.0	94.0	13.3	86.7	18.0	82.0	5.3	94.7
2002	14.6	85.4	20.7	79.3	5.1	94.9	12.2	87.8	16.9	83.1	4.4	95.6
2003	15.0	85.0	21.2	78.8	5.7	94.3	12.6	87.4	17.3	82.7	4.9	95.1
2004	14.8	85.2	21.5	78.5	5.4	94.6	12.6	87.4	17.3	82.7	4.9	95.1
2005	14.2	85.8	20.5	79.5	5.0	95.0	12.0	88.0	16.6	83.4	4.5	95.5
2006	12.9	87.1	20.1	79.9	3.9	96.1	11.7	88.3	16.2	83.8	4.1	95.9
2007	13.9	86.1	20.2	79.8	5.4	94.6	12.1	87.9	16.3	83.7	5.1	94.9
2008	13.9	86.1	21.5	78.5	4.3	95.7	12.4	87.6	17.2	82.8	4.3	95.7

*Includes soft drinks and alcoholic beverages. USDA, Economic Research Service.

When BLS reports its annual indexes of consumer food prices and producer commodity prices, its measure explicitly controls for product-mix changes. By using a fixed basket of food products, the BLS index of prices avoids factors such as the introduction of marinated beef or a shift from limited-service to full-service restaurants. The importance of price comparisons under the two approaches is demonstrated in figure 3, which reports the BLS annual indexes of consumer food prices for at-home and food-away expenditures from 1993 to 2008. The figure also reports a total farm commodity (food and nonfood) producer price index (PPI) over the same interval. These price indexes support the IO analysis finding that the farm share of food dollar expenditures increased substantially in 2007 and 2008 due to higher farm commodity prices; however, the BLS price indexes also show very little difference in the year-to-year changes to at-home versus food-away food prices. The BLS data appear to contrast with the results in table 1 that indicate that the food-away marketing bill share of the food dollar increased throughout most of the study period while the food-at-home-share remained roughly constant.

The divergence between the BLS data and the ERS data in table 1 can be explained by several possible trends. If the bundle of food-away meals purchased by U.S. food consumers changed during this period, with consumers buying more food and/or services with higher marketing margins, this change would be refl ected in the farm share measure but would not immediately show up in the fixed-bundle price indexes. In addition, farm commodity producer prices paid by the foodservice industry may have declined relative to the prices paid by food retailers, which can occur in the absence of overall changes to the price of farm commodities. To examine this issue more closely, a new approach to measuring the components of food dollar expenditures is introduced in the next section.

COMPONENTS OF THE FOOD DOLLAR

In the next step, we use supply chain analysis to trace the market value of total food dollar expenditures back to the sources of value added and to the assets employed by participants in the food dollar supply chain.

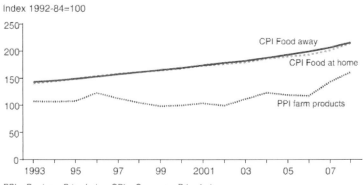

PPI = Producer Price Index; CPI = Consumer Price Index.
Source: Bureau of Labor Statistics, U.S. Department of Labor.
Figure 3. Index of farm and food commodity prices, 1993-08.

Within the IO accounting framework, the market value of all final market sales, including all food dollar expenditures, is exactly equal to the sum total of all value added by every industry that is either directly or indirectly linked to the commodity supply chain. This result is simple to demonstrate by recalling that equation 1 identifies the gross outputs of all industries directly and indirectly linked to the import-inclusive food dollar expenditures. Multiplying both sides of equation 1 by the inverse of the Leontief matrix, and noting that inverting the Leontief matrix and summing the resulting matrix down each column produces a row vector of industry value-added coefficients, v_m', gives the following:[8]

$$\underbrace{v\_m'_C \cdot x_C^{fd}}_{\substack{\text{summation of} \\ \text{value added to food dollar} \\ \text{across all C industries}}} = \underbrace{i'_C \cdot y_C^{fd}}_{\substack{\text{summation of} \\ \text{import-inclusive food} \\ \text{dollar expenditures}}}$$

(4)

In this expression, the a subscript used in equation 1 was replaced by the C subscript, since all industry contributions to market value are studied in the present context, not just the farm share. Although equation 4 precisely gives the value contribution of all industries supporting the food dollar supply chain, this turns out to be too much information; most of the roughly 400 industries in the BEA detailed benchmark IO accounts, and roughly 200 industries in the annual BLS IO accounts, either directly or indirectly contribute some value to the production of food. The challenge is how to process this information in a way that informs our understanding of how value accumulates along the supply chain, from the production and application of farm inputs to the purchase of food products by or for U.S. households, as well as to show how this process changes from year to year.

A matrix reduction procedure in IO analysis (see chapter 3 in Leontief, 1986) to facilitate supply chain studies is suited for precisely this type of problem. Developed in the 1960s to facilitate a supply chain study of the U.S. metalworking industries, the original application identifi ed four branches of production belonging to the metalworking industry group. Using precise mathematical computations, all industries <u>not</u> identifi ed with these branches of production—here we use the term supply chain (SC) industries—were eliminated, but their value-added contributions were exactly allocated to the SC industries in proportion to the materials and services supplied. It is useful to refer to these non-SC industries as subcontracting establishments.

A simple aggregation of key industry groups that comprise the food supply chain industries can provide direct measurements of the value added to the food dollar by each group. A substantial portion of the value from each food dollar expenditure, however, is produced by numerous other industries that support some or all of the identifi ed supply chain industries. The wholesale trade industry is a case in point. Agribusiness wholesalers, grocery wholesalers, and foodservice wholesalers provide supplies and services to three very distinct food supply chain groups. The role and structure of each of these three wholesale industry groups has evolved differently over time, and the groups are more closely allied to their industry clients than to each other. The matrix reduction procedure introduced in this section is a systematic approach to measure these relationships, whereas a simple aggregation of the wholesale trade industry would obscure the relationships.

To state this measure formally, let S denote the set of M different supply chain industry groups that facilitate production and delivery of food commodities to U.S. households, such that $S = \{s_1, s_2, ..., s_M\}$. Let v_m^* denote the reestimated value-added coeffi cients representing only the supply chain industries, but also refl ecting the combined value-added coefficients of their subcontracting industries. Equation 4 is then modifi ed for the reduced food supply chain JO system and to refl ect the deduction of household import food expenditures, as follows:

$$v\_m_S^{*'} \cdot x_S^{fd\_net} = y_S^{fd\_net},$$

$$where: \quad x_S^{fd\_net} = x_S^{fd} - \overline{S\_m}_S \cdot y_S^{fd},$$

$$and: \quad y_S^{fd\_net} = (i_S - S\_m_S)' \cdot y_S^{fd}. \tag{5}$$

Each product of a supply chain value-added coeffi cient and its corresponding net industry output represents the value contribution of the specifi c supply chain industry. Dividing each element-wise product by the summation on the right in equation 5 produces the value contributions of each supply chain industry, and its subcontractors, to each food dollar expenditure:

$$industry\ group\ value - added\ food\ dollar = \overline{v\_m}_S^{*} \cdot x_S^{fd\_net} \div y_S^{fd\_net} \tag{6}$$

Industry Group Value-Added Composition of the Food Dollar

To carry out the supply chain analysis, supply chain industries are clustered into 10 industry groups, based on their contributions to the different stages of food production or to key food supply chain services.

The following supply chain industry groupings were selected:
1) Farm and agribusiness
2) Food processing
3) Food retailing
4) Foodservices (restaurants and other establishments serving food away from home)
5) Transportation
6) Energy
7) Packaging
8) Finance and insurance
9) Advertising
10) Legal, accounting, and bookkeeping

Figure 4 summarizes the value-added components of the 2006 food dollar by industry group, as formally stated in equation 6, to facilitate a comparison with 2006 cost-component measures from the old food marketing bill series reported in figure 1. Under the new JO-based food dollar series, a complete accounting of each supply chain industry group's contribution to the value of food purchases is measured and reported. This facilitates a more informative account of the roles and impacts of the different industry groups in the formation of food market values and the effects of the industries on producer prices of food commodities. For example, consider the segments labeled "Energy" in figures 1 and 4. Under the old food marketing bill series (fi gure 1), the energy segment represented the average costs per food dollar expenditure for electricity, natural gas, and other fuel purchases by food processing, wholesaling, retailing, and foodservice establishments. Energy costs of the farm and agribusiness, transportation, and packaging industries, for example, are not refl ected in the 3.5-cent energy segment in this figure. In figure 4, the same 2006 food dollar reported 5.8 cents of energy per food dollar expenditure, because this larger measure incorporates the energy value of every food dollar supply chain industry. This is part of the reason the transportation and packaging industry groups show smaller segments in figure 4 than in figure 1. Both transportation and packaging are energy-intensive industries, and when the value of energy used by these groups is deducted from their contribution to the food dollar, their contributions noticeably decline. The same reasoning applies to each industry group reported in figure 4. For example, any packaging, fi nance and insurance, and transportation service costs incurred by each of the other supply chain industry groups are not refl ected in the value contributions of those industries, but are instead consolidated into the appropriate industry group segment.

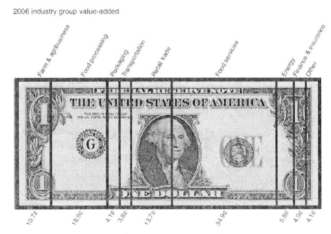

Figure 4. (Continued).

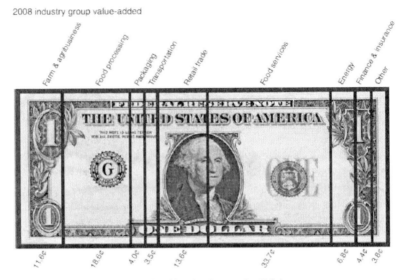

Source: USDA, Economic Research Service.

Figure 4. Industry group value-added shares of the food dollar.

Another informative comparison is between the farm share segment in figure 2 and the farm and agribusiness value-added segment in figure 4. In figure 4, farm and agribusiness value-added contributions to the food dollar that trace back to other supply chain industry groups, such as energy and transportation, are deducted from the farm share value reported in figure 2 in order to arrive at the farm and agribusiness value-added contribution. For 2006, these energy, transportation, and other services amounted to 3.5 cents (14.2 - 10.7) of each food dollar. The remaining 10.7 cents represents value contributions of the industry assets employed by farm and agribusiness establishments to support the production of farm commodities.

Primary Factor Return Composition of the Food Dollar

The next step is to distribute industry value-added proceeds to the owners of the primary factors employed by industry.

Each industry produces a product by a process that typically involves purchasing output products of other industries (intermediate inputs) and then employing its own industry assets to use these intermediate inputs to produce a new and different output. The industry assets are the primary factors of production that add value to the intermediate products purchased from other industries. In IO accounts, industry value added is recorded as income and is allocated as returns to primary factors as follows (type of payments to the factor owners in parenthesis):

a) Domestic hired labor (salary and benefi ts)
b) Domestic industry capital (returns to capital;[9] taxes on output[10])
c) International assets[11] (international payments)

With the exception of imports, the BLS annual IO accounts do not itemize industry value added, but report the combined value added by food supply chain industries. BEA annual IO accounts provide this detailed breakout beginning in 1997, and the information is incorporated for the available years (see appendix for detailed explanation).

BEA annual data are used to measure the primary factor value-added food dollar, obtained by a restatement of equation 6 with the addition of a second subscript, p, defi ned over each type of primary factor v_mS,P (see equation A35 in the appendix for a formal statement of this measure):

$$primary\ factor\ value\ added\ food\ dollar = v\_m^*_{P,S} \cdot x_S^{fd\_net} \div y_S^{fd\_net} \tag{7}$$

Figure 5 summarizes the allocation of the 2006 food dollar into total payments to primary factors of production, obtained from application of equation 7 to the 2006 IO data. A comparison of the salary and benefi ts segments in figure 5 with the labor segment of the old food marketing bill series (fi gure 1) again reveals a stark contrast. In the present analysis (fi g. 5), salary and benefi ts include all domestic labor compensation for work toward fulfi lling the demand for all food dollar purchases. Figure 5 indicates that slightly over half of every food dollar expenditure covers salaries and benefi ts of U.S. workers. Previously, labor excluded farm and agribusiness labor earnings (fi g. 1), since only food marketing bill cost components were measured. The old series also had gaps in the measurement of labor earnings in other supply chain industries, including those for some transportation and manufacturing industry workers. These differences resulted in the substantially lower reported labor earnings per food dollar in figure 1.

Analysis of Industry Group Primary Factor Returns:
1993 to 2008

The next steps measure changes to the value composition of food dollar expenditures over time to assess changes to the structure of the food system and the composition of food expenditures.

A more informative allocation of the food dollar market value is by a cross-tabulation of industry group value added by the primary factor group. This is obtained by conversion of the net industry output vector in equation 7 to a diagonal matrix:

$$cross - tabulation\ value\ added\ food\ dollar = v\_m_{P,S}^* \cdot \overline{x}_S^{fd\_net} \div y_S^{fd\_net} \quad (8)$$

Table 3 reports the computation of equation 8 using data on U.S. total, at-home, and food-away nominal food dollar expenditures from 1993 to 2008. Results are reported for each primary factor return and total return to assets and for all industries combined. Results are also itemized for each of the 10 supply chain industry groups, for a total of 165 categories. Table lines are numbered for ease of reference.

Overall, the salaries of U.S. workers accounted for the highest portion of the food dollar among primary factor returns (table 3, line 2). Returns to workers comprise wages, salaries, and benefi ts, such as both employer and employee contributions to pension and insurance funds, and include all income and wage taxes (State and Federal) on these salaries. In 1997, the salaries of U.S. workers accounted for half of total food dollar expenditures. This amount climbed 10 percent by 2001, reaching about 55 cents of every food dollar spent, on average, but had fallen back to 51 cents by 2008. Among supply chain industry groups, foodservice workers accounted for the largest share of salary and benefi ts, reaching nearly 22 cents of the total food dollar in 2006 (table 3, line 32). Most of the supply chain industry groups mirrored the overall trend of increasing labor returns from the late 1990s into the early 2000s, followed by a dropping off through 2008. A notable exception to this trend was farming and agribusiness labor, where labor returns per food dollar expenditure generally declined throughout the study period, from a high of about 4 cents in 1997 to a low of 2.8 cents in 2007 (table 3, line 7).

Industry capital, or property, has the next highest primary factor returns per food dollar expenditure, ranging between 32 and 36 cents for the average dollar spent on food between 1997 and 2008 (table 3, line 3). Property incomes were at their highest share (35.8 cents) of overall food dollar expenditures in 1997, and at their lowest in 2006 at just under 32 cents. Property incomes amounted to roughly one-third of the total food dollar in 2007 and 2008. Farm and agribusiness property income (including land) were highest among food supply chain industry groups from 1997 to 2001, reaching 8.3 cents in 1999 (table 3, line 8). From 2002 onward, the foodservice industry had the highest returns to capital, surpassing 9 cents of every dollar spent on U.S.-produced food by U.S. households in 2006 and 2008 (table 3, line 33). Property income of food processors fell off substantially over the study period (table 3, line 13). On a percentage basis, property income for packaging and advertising also dropped substantially, while returns to capital for the energy and the legal/accounting/bookkeeping industries trended higher throughout the study period as a share of total food dollar expenditures.

Taxes (less subsidies) and fees on industry output include Federal excise taxes and custom duties, State and local sales and property taxes, and some nontax fees and assessments. From 1997 to 2008, these production costs ranged between a low of 6.7 cents in 2000 to a high of 9.0 cents in 1997 (table 3, row 4). The highest property income taxes fell on the food retailing and foodservice industries (table 3, rows 29 and 34). Overall output taxes on farming and agribusiness were negative in some years, owing to the Federal production subsidies on numerous farm commodities (table 3, line 9).

Imports of food and nonfood ingredients have been the fastest growing of the primary factor returns. Because detailed country-of-origin input-output tables are not available for these imports, only the total primary factor value contributions are reported for each supply chain industry group, based on the commodity values of imports entering the country.[12] Less than 5 cents of the value of U.S.-produced food sold to U.S. households in 1993 was for imported commodities. By 2008, this value had increased 65 percent, reaching 7.8 cents of each dollar spent on U.S.-produced foods (table 3, line 5). Between 1993 and 2001, farm and agribusiness imports were the leading contributors to the import value of U.S.-produced food (table 3, line 10). In 2002, the energy industry overtook all other industries in total import value contributions to the food dollar, and by 2008 imported energy of 2.5 cents on the dollar (table 3, line 40) was almost twice that of the industry group with the next highest import value, farm and agribusiness at 1.4 cents (table 3, line 10). The overall increase in import share of the food dollar between 2001 and 2008 coincides with declines in the domestic wage share over this period.

Table 3. Cost components of the U.S. nominal food dollar, 1993 to 2008

Primary factor cost	1990s							2000s								
	93	94	95	96	97	98	99	2000	01	02	03	04	05	06	07	08
								Total food dollar								
All industries																
1. Total	100.0	100.0	100.0	100.0	100.0	100.0	100.0	100.0	100.0	100.0	100.0	100.0	100.0	100.0	100.0	100.0
2. Labor					50.0	53.9	53.8	54.6	54.9	54.7	53.4	53.4	52.9	53.2	51.3	50.8
3. Property					35.8	33.3	34.0	32.6	32.8	32.3	33.5	32.7	32.8	31.9	33.2	33.0
4. Taxes					9.0	7.6	6.8	6.7	6.7	7.6	7.4	7.7	7.3	7.9	8.1	8.4
5. Imports	4.7	4.6	5.3	5.3	5.1	5.2	5.4	6.0	5.6	5.5	5.7	6.2	7.0	7.0	7.4	7.8
Farm and agribusiness																
6. Total	14.5	14.0	14.4	14.1	14.1	13.5	12.9	12.4	12.0	11.8	11.9	11.6	11.3	10.7	11.8	11.6
7. Labor					4.0	3.9	4.0	3.9	3.8	3.9	3.4	3.0	3.0	3.0	2.8	2.9
8. Property					7.6	8.2	8.3	8.0	7.6	6.8	7.6	7.3	7.4	6.5	7.3	7.0
9. Taxes					1.1	0.0	-0.7	-0.8	-0.6	0.1	-0.2	0.2	-0.4	0.0	0.4	0.3
10. Imports	1.3	1.2	1.5	1.4	1.5	1.4	1.4	1.3	1.1	1.1	1.1	1.1	1.2	1.2	1.3	1.4
Food processing																
11. Total	21.7	22.6	22.5	23.4	23.6	23.4	22.7	21.7	21.6	21.3	20.9	20.3	20.4	18.9	19.0	18.6
12. Labor					11.8	13.9	13.0	13.0	12.7	12.7	12.2	12.5	12.6	11.8	11.2	10.9
13. Property					9.1	7.2	7.5	6.6	6.9	6.7	6.7	5.8	5.7	5.2	5.8	5.6
14. Taxes					1.5	1.1	1.0	1.0	0.9	1.0	1.0	0.9	1.0	0.9	0.9	0.9
15. Imports	1.1	1.1	1.2	1.3	1.2	1.2	1.2	1.2	1.1	1.0	1.0	1.1	1.1	1.0	1.2	1.2
Packaging																
16. Total	4.7	4.8	5.2	5.0	4.7	4.8	4.8	4.7	4.7	4.7	4.5	4.3	4.3	4.1	4.1	4.0
17. Labor					2.3	2.3	2.3	2.3	2.4	2.3	2.3	2.0	1.9	1.8	1.8	1.8
18. Property					1.3	1.3	1.3	1.3	1.2	1.2	1.1	1.2	1.1	1.1	1.0	1.0
19. Taxes					0.2	0.2	0.2	0.2	0.2	0.2	0.2	0.1	0.2	0.1	0.1	0.1
20. Imports	0.8	0.8	1.0	1.0	0.8	0.9	1.0	1.0	1.0	1.0	1.0	1.1	1.1	1.1	1.1	1.1

Transportation

21.Total	4.5	4.6	4.5	4.5	4.4	4.4	4.3	4.1	4.0	4.0	3.8	3.8	3.8	3.8	3.5	3.5
22. Labor					2.5	2.7	2.6	2.5	2.5	2.4	2.3	2.2	2.2	2.1	1.9	2.0
23. Property					1.4	1.4	1.3	1.2	1.2	1.2	1.2	1.2	1.3	1.3	1.2	1.1
24.Taxes					0.2	0.2	0.2	0.2	0.1	0.1	0.1	0.1	0.1	0.1	0.1	0.1
25. Imports	0.2	0.2	0.2	0.2	0.3	0.2	0.3	0.3	0.2	0.2	0.2	0.2	0.2	0.2	0.2	0.2

Retail trade

26. Total	13.9	14.3	14.4	14.7	14.6	14.4	14.5	14.2	14.2	14.4	14.6	14.0	14.3	13.7	13.8	13.6
27. Labor					8.3	8.0	8.1	8.1	8.1	8.2	8.1	7.6	7.7	7.3	7.4	7.5
28. Property					3.8	3.4	3.4	3.1	3.1	3.3	3.6	3.4	3.6	3.5	3.5	3.2
29. Taxes					2.2	2.7	2.7	2.7	2.6	2.6	2.7	2.6	2.7	2.7	2.7	2.6
30. Imports	0.2	0.2	0.2	0.3	0.2	0.3	0.3	0.3	0.3	0.2	0.3	0.3	0.3	0.3	0.3	0.3

Foodservice

31.Total	28.3	27.9	27.5	26.6	27.0	27.6	28.5	29.4	30.0	30.4	30.9	32.3	31.5	34.9	33.3	33.7
32. Labor					16.3	17.7	18.1	18.8	19.3	19.3	19.5	20.4	19.8	21.8	20.6	20.3
33. Property					7.5	6.8	7.1	7.2	7.4	7.8	8.0	8.4	8.1	9.1	8.8	9.2
34. Taxes					3.0	2.6	2.7	2.8	2.8	2.8	2.9	3.0	3.0	3.3	3.1	3.4
35 Imports	0.4	0.4	0.4	0.5	0.3	0.5	0.6	0.6	0.6	0.5	0.5	0.6	0.6	0.7	0.7	0.8

Table 3. (Continued)

Primary factor cost	1990s							2000s								
	93	94	95	96	97	98	99	2000	01	02	03	04	05	06	07	08
								Total food dollar								
Energy																
36. Total	5.0	4.6	4.3	4.4	4.2	3.9	4.1	4.8	5.0	4.9	5.1	5.2	6.2	5.8	6.2	6.8
37. Labor					1.1	1.1	1.2	1.2	1.3	1.3	1.1	1.1	1.2	1.0	1.1	1.3
38. Property					1.9	1.9	1.9	2.1	2.2	2.0	2.1	2.1	2.4	2.1	2.3	2.3
39. Taxes					0.5	0.4	0.4	0.4	0.4	0.5	0.4	0.4	0.4	0.4	0.4	0.6
40. Imports	0.6	0.6	0.6	0.6	0.7	0.4	0.6	1.0	1.0	1.2	1.4	1.7	2.2	2.2	2.3	2.5
Finance and insurance																
41. Total	3.7	3.6	3.5	3.6	3.7	3.9	4.1	4.3	4.3	4.2	4.0	4.1	4.0	3.9	4.1	4.3
42. Labor					1.7	2.1	2.3	2.4	2.3	2.3	2.2	2.2	2.2	2.1	2.3	2.3
43. Property					1.7	1.6	1.6	1.7	1.7	1.8	1.7	1.7	1.7	1.6	1.7	1.8
44. Taxes					0.2	0.1	0.1	0.1	0.1	0.1	0.1	0.1	0.1	0.1	0.1	0.2
45. Imports	0.0	0.0	0.0	0.1	0.1	0.1	0.1	0.1	0.1	0.1	0.1	0.1	0.1	0.1	0.1	0.1
Advertising																
46. Total	2.4	2.3	2.4	2.5	2.6	2.7	2.7	2.7	2.5	2.4	2.4	2.4	2.3	2.3	2.3	2.0
47. Labor					1.4	1.5	1.5	1.6	1.5	1.4	1.3	1.3	1.3	1.3	1.2	1.1
48. Property					1.0	1.0	1.0	0.9	0.9	0.9	0.9	1.0	0.9	0.9	0.9	0.8
49. Taxes					0.1	0.1	0.1	0.1	0.1	0.1	0.1	0.1	0.1	0.1	0.1	0.1
50. Imports	0.1	0.1	0.1	0.1	0.1	0.1	0.1	0.1	0.1	0.1	0.1	0.1	0.1	0.1	0.1	0.1

	C1	C2	C3	C4	C5	C6	C7	C8	C9	C10	C11	C12	C13	C14	C15	C16
Legal, accounting, and bookkeeping																
51. Total	1.3	1.3	1.3	1.3	1.3	1.4	1.5	1.6	1.7	1.8	1.8	1.8	1.8	1.8	1.7	1.8
52. Labor					0.8	0.9	0.9	1.0	1.1	1.1	1.1	1.1	1.1	1.1	1.0	1.1
53. Property					0.5	0.5	0.5	0.5	0.6	0.6	0.6	0.7	0.7	0.6	0.6	0.7
54. Taxes					0.0	0.0	0.0	0.0	0.1	0.1	0.1	0.1	0.1	0.1	0.1	0.1
55. Imports	0.0	0.0	0.0	0.0	0.0	0.0	0.0	0.0	0.0	0.0	0.0	0.0	0.0	0.0	0.0	0.0

At-home food dollar

	C1	C2	C3	C4	C5	C6	C7	C8	C9	C10	C11	C12	C13	C14	C15	C16
All industries																
56. Total	100.0	100.0	100.0	100.0	100.0	100.0	100.0	100.0	100.0	100.0	100.0	100.0	100.0	100.0	100.0	100.0
57. Labor					46.8	50.5	50.3	51.3	51.3	51.3	49.6	49.5	49.0	49.1	46.6	46.2
58. Property					38.3	36.0	37.2	35.6	35.9	34.8	36.6	35.5	35.7	34.7	36.2	36.2
59. Taxes					8.7	7.3	6.1	6.0	6.0	7.2	6.9	7.4	6.7	7.5	7.9	7.8
60. Imports	5.6	5.5	6.2	6.2	6.2	6.2	6.4	7.1	6.7	6.6	7.0	7.6	8.6	8.8	9.3	9.8
Farm and agribusiness																
61. Total	19.8	19.0	19.5	18.7	18.9	18.5	18.1	17.7	17.4	17.4	17.6	17.6	17.0	17.2	18.5	18.3
62. Labor					5.4	5.3	5.6	5.5	5.6	5.7	5.0	4.6	4.6	4.8	4.4	4.3
63. Property					10.2	11.3	11.7	11.6	11.2	10.1	11.3	11.2	11.3	10.5	11.5	11.3
64. Taxes					1.4	0.0	-1.1	-1.2	-1.0	0.1	-0.3	0.2	-0.7	0.0	0.5	0.5
65. Imports	1.8	1.6	2.0	1.8	1.9	1.9	1.9	1.8	1.6	1.5	1.6	1.6	1.8	1.9	2.1	2.1
Food processing																
66. Total	30.7	31.8	31.3	32.1	32.5	32.6	32.5	31.9	32.1	32.1	31.7	31.9	31.5	31.3	30.5	30.1
67. Labor					16.2	19.3	18.6	19.1	18.8	19.0	18.5	19.6	19.4	19.5	17.9	17.7
68. Property					12.6	10.1	10.8	9.8	10.4	10.1	10.2	9.1	8.8	8.6	9.3	9.1
69. Taxes					2.1	1.5	1.4	1.4	1.4	1.4	1.5	1.5	1.5	1.5	1.5	1.4
70. Imports	1.5	1.6	1.6	1.7	1.6	1.7	1.7	1.7	1.6	1.5	1.6	1.7	1.7	1.7	1.9	1.8

Table 3. (Continued)

Primary factor cost	1990s							2000s								
	93	94	95	96	97	98	99	2000	01	02	03	04	05	06	07	08
At-home food dollar																
Packaging																
71. Total	5.6	5.7	6.1	5.9	5.5	5.7	5.7	5.7	5.7	5.8	5.5	5.3	5.3	5.1	5.0	4.9
72. Labor					2.7	2.8	2.8	2.7	2.9	2.9	2.8	2.4	2.4	2.2	2.2	2.2
73. Property					1.5	1.6	1.6	1.5	1.4	1.5	1.3	1.4	1.4	1.3	1.2	1.2
74. Taxes					0.2	0.2	0.2	0.2	0.2	0.2	0.2	0.2	0.2	0.2	0.2	0.2
75. Imports	0.9	1.0	1.2	1.1	1.0	1.1	1.2	1.3	1.2	1.2	1.2	1.3	1.3	1.4	1.3	1.4
Transportation																
76. Total	5.8	5.9	5.8	5.7	5.6	5.8	5.7	5.5	5.5	5.5	5.3	5.5	5.4	5.6	5.1	5.1
77. Labor					3.2	3.4	3.4	3.3	3.4	3.3	3.1	3.2	3.1	3.1	2.8	2.8
78. Property					1.8	1.8	1.7	1.6	1.6	1.7	1.7	1.8	1.8	1.9	1.8	1.8
79. Taxes					0.3	0.2	0.2	0.2	0.2	0.2	0.2	0.2	0.2	0.2	0.2	0.2
80. Imports	0.2	0.2	0.2	0.3	0.4	0.3	0.4	0.4	0.3	0.3	0.3	0.3	0.3	0.3	0.4	0.4
Retail trade																
81. Total	24.1	24.3	24.2	24.2	24.2	24.0	24.1	23.9	23.9	24.0	24.7	24.3	24.4	24.8	24.3	24.2
82. Labor					13.8	13.3	13.4	13.7	13.7	13.6	13.7	13.3	13.1	13.2	12.9	12.8
83. Property					6.3	5.8	5.7	5.2	5.3	5.6	6.0	6.0	6.1	6.3	6.2	6.2
84. Taxes					3.7	4.5	4.5	4.5	4.4	4.4	4.5	4.5	4.6	4.8	4.7	4.6
85. Imports	0.3	0.3	0.4	0.4	0.4	0.5	0.5	0.6	0.5	0.4	0.5	0.5	0.5	0.5	0.6	0.6
Foodservice																
86. Total	0.7	0.7	0.8	0.8	0.8	0.8	0.8	0.9	0.9	1.0	1.0	1.0	0.9	0.9	0.9	0.9
87. Labor					0.5	0.5	0.5	0.6	0.6	0.6	0.6	0.6	0.6	0.6	0.5	0.5
88. Property					0.3	0.2	0.2	0.2	0.3	0.3	0.3	0.3	0.3	0.3	0.3	0.3
89. Taxes					0.1	0.1	0.1	0.1	0.1	0.1	0.1	0.1	0.1	0.1	0.1	0.1
90. Imports	0.0	0.0	0.0	0.0	0.0	0.0	0.0	0.0	0.0	0.0	0.0	0.0	0.0	0.0	0.0	0.0

Energy																
91. Total	5.7	5.1	4.8	5.0	4.7	4.3	4.6	5.5	5.7	5.7	5.9	6.2	7.4	6.9	7.5	8.2
92. Labor					1.2	1.2	1.3	1.4	1.5	1.4	1.3	1.2	1.3	1.2	1.3	1.4
93. Property					2.1	2.1	2.2	2.4	2.5	2.3	2.4	2.4	2.8	2.5	2.7	3.1
94. Taxes					0.5	0.5	0.5	0.5	0.5	0.5	0.5	0.5	0.5	0.5	0.5	0.6
95. Imports	0.8	0.7	0.7	0.8	0.9	0.5	0.7	1.2	1.3	1.5	1.7	2.1	2.7	2.8	3.0	3.2
Finance and Insurance																
96. Total	3.9	3.8	3.6	3.7	3.8	4.0	4.3	4.6	4.6	4.6	4.3	4.4	4.3	4.3	4.5	4.8
97. Labor					1.7	2.2	2.4	2.5	2.5	2.4	2.3	2.4	2.3	2.3	2.4	2.5
98. Property					1.8	1.7	1.7	1.8	1.9	1.9	1.8	1.8	1.8	1.8	1.8	2.0
99. Taxes					0.2	0.1	0.2	0.2	0.2	0.2	0.1	0.2	0.2	0.1	0.2	0.2
100. Imports	0.0	0.0	0.0	0.1	0.1	0.1	0.1	0.1	0.1	0.1	0.1	0.1	0.1	0.1	0.1	0.1
Advertising																
101. Total	2.5	2.4	2.5	2.6	2.7	2.7	2.7	2.7	2.4	2.3	2.3	2.3	2.2	2.2	2.1	1.9
102. Labor					1.5	1.6	1.5	1.6	1.4	1.3	1.3	1.2	1.2	1.2	1.2	1.0
103. Property					1.1	1.0	1.0	0.9	0.8	0.9	0.9	0.9	0.9	0.9	0.8	0.8
104. Taxes					0.1	0.1	0.1	0.1	0.1	0.1	0.1	0.1	0.1	0.1	0.1	0.1
105. Imports	0.1	0.1	0.1	0.1	0.1	0.1	0.1	0.1	0.1	0.1	0.1	0.1	0.1	0.1	0.1	0.1

Table 3. (Continued)

Primary factor cost	1990s							2000s								
	93	94	95	96	97	98	99	2000	01	02	03	04	05	06	07	08
At-home food dollar																
Legal, accounting, and bookkeeping																
106. Total	1.4	1.3	1.4	1.4	1.4	1.5	1.5	1.6	1.7	1.7	1.8	1.7	1.7	1.7	1.7	1.8
107. Labor					0.8	0.9	0.9	1.0	1.1	1.1	1.1	1.0	1.0	1.0	1.0	1.0
108. Property					0.6	0.5	0.5	0.5	0.6	0.6	0.6	0.6	0.6	0.6	0.6	0.6
109. Taxes					0.0	0.0	0.0	0.0	0.1	0.1	0.1	0.1	0.1	0.1	0.1	0.1
110. Imports	0.0	0.0	0.0	0.0	0.0	0.0	0.0	0.0	0.0	0.0	0.0	0.0	0.0	0.0	0.0	0.0
Away-from-home food dollar																
All industries																
111. Total	100.0	100.0	100.0	100.0	100.0	100.0	100.0	100.0	100.0	100.0	100.0	100.0	100.0	100.0	100.0	100.0
112. Labor					54.6	58.6	58.7	59.3	59.8	59.6	58.8	58.5	58.2	58.0	57.1	56.9
113. Property					32.3	29.5	29.4	28.4	28.4	28.6	29.2	29.0	28.9	28.8	29.5	29.5
114. Taxes					9.6	7.9	7.8	7.8	7.7	8.0	8.0	8.1	8.1	8.3	8.2	8.2
115. Imports	3.7	3.7	4.1	4.2	3.6	4.0	4.1	4.5	4.1	3.8	4.0	4.3	4.8	4.9	5.2	5.4

Farm and agribusiness

116. Total	8.0	7.5	7.5	7.3	7.2	6.3	5.4	4.8	4.2	3.6	3.9	3.7	3.6	3.1	3.6	3.4
117. Labor					2.0	1.8	1.6	1.4	1.3	1.2	1.1	1.0	1.0	0.9	0.9	0.8
118. Property					3.8	3.7	3.3	2.9	2.5	2.0	2.3	2.2	2.3	1.8	2.1	2.0
119. Taxes					0.6	0.1	-0.2	-0.2	-0.1	0.1	0.0	0.1	-0.1	0.1	0.1	0.1
120. Imports	0.9	0.8	0.9	0.8	0.9	0.8	0.7	0.6	0.5	0.4	0.4	0.4	0.5	0.4	0.5	0.5

Food processing

121. Total	10.3	10.3	10.5	10.8	10.8	10.1	8.4	7.1	6.3	5.7	5.7	5.2	5.3	4.4	4.8	4.6
122. Labor					5.5	6.0	4.8	4.3	3.8	3.4	3.4	3.2	3.3	2.7	2.8	2.7
123. Property					3.9	3.0	2.6	2.1	1.9	1.7	1.8	1.4	1.4	1.2	1.4	1.4
124. Taxes					0.7	0.5	0.4	0.3	0.3	0.3	0.3	0.2	0.3	0.2	0.2	0.2
125. Imports	0.7	0.7	0.7	0.7	0.6	0.6	0.5	0.5	0.4	0.3	0.3	0.3	0.4	0.3	0.4	0.4

Packaging

126. Total	3.5	3.6	3.9	3.7	3.5	3.5	3.4	3.4	3.2	3.2	3.1	3.0	3.0	2.9	2.9	2.9
127. Labor					1.7	1.7	1.6	1.6	1.6	1.6	1.5	1.3	1.3	1.2	1.3	1.2
128. Property					1.0	1.0	1.0	0.9	0.8	0.8	0.8	0.8	0.8	0.7	0.7	0.7
129. Taxes					0.2	0.1	0.1	0.1	0.1	0.1	0.1	0.1	0.1	0.1	0.1	0.1
130. Imports	0.6	0.6	0.8	0.7	0.6	0.7	0.7	0.8	0.7	0.7	0.7	0.8	0.8	0.8	0.8	0.8

Transportation

131. Total	2.8	2.8	2.7	2.7	2.6	2.5	2.3	2.1	1.9	1.8	1.8	1.8	1.7	1.7	1.6	1.5
132. Labor					1.5	1.5	1.3	1.2	1.2	1.1	1.1	1.0	1.0	0.9	0.8	0.8
133. Property					0.8	0.8	0.7	0.6	0.6	0.5	0.6	0.6	0.6	0.6	0.5	0.5
134. Taxes					0.1	0.1	0.1	0.1	0.1	0.1	0.1	0.1	0.1	0.1	0.1	0.1
135. Imports	0.1	0.1	0.1	0.1	0.2	0.2	0.2	0.2	0.1	0.1	0.1	0.1	0.1	0.1	0.1	0.1

Retail trade

136. Total	0.7	0.8	0.8	0.8	0.7	0.6	0.7	0.6	0.4	0.5	0.5	0.5	0.6	0.6	0.6	0.5
137. Labor					0.4	0.4	0.4	0.3	0.2	0.3	0.3	0.3	0.3	0.3	0.3	0.3
138. Property					0.2	0.2	0.2	0.1	0.1	0.1	0.1	0.1	0.1	0.2	0.2	0.1
139. Taxes					0.1	0.1	0.1	0.1	0.1	0.1	0.1	0.1	0.1	0.1	0.1	0.1
140. Imports	0.0	0.0	0.0	0.0	0.0	0.0	0.0	0.0	0.0	0.0	0.0	0.0	0.0	0.0	0.0	0.0

Table 3. (Continued)

Primary factor cost	1990s							2000s								
	93	94	95	96	97	98	99	2000	01	02	03	04	05	06	07	08
Away-from-home food dollar																
Foodservice																
141. Total	63.8	64.5	64.3	64.3	64.6	66.1	68.5	69.9	71.8	73.1	73.0	73.7	73.1	75.0	73.9	74.1
142. Labor					39.1	42.1	43.5	44.5	45.9	46.4	46.0	46.3	45.9	46.7	45.7	45.8
143. Property					17.9	16.4	17.0	17.1	17.7	18.7	18.9	19.0	18.7	19.6	19.6	19.7
144. Taxes					7.1	6.4	6.6	6.7	6.7	6.8	6.8	6.9	7.0	7.2	6.9	6.9
145. Imports	0.9	1.0	1.1	1.2	0.6	1.3	1.4	1.6	1.5	1.3	1.3	1.4	1.5	1.6	1.7	1.8
Energy																
146. Total	4.2	3.8	3.7	3.7	3.5	3.3	3.4	3.8	3.9	3.8	3.9	4.0	4.6	4.4	4.7	5.0
147. Labor					0.9	1.0	1.0	1.0	1.0	1.0	0.9	0.9	0.9	0.8	0.9	0.9
148. Property					1.6	1.7	1.7	1.8	1.8	1.6	1.7	1.7	1.9	1.7	1.8	2.0
149. Taxes					0.4	0.4	0.4	0.4	0.3	0.4	0.4	0.3	0.4	0.4	0.4	0.4
150. Imports	0.5	0.4	0.4	0.5	0.6	0.3	0.4	0.7	0.7	0.8	1.0	1.1	1.5	1.5	1.6	1.7
Finance and insurance																
151. Total	3.5	3.4	3.3	3.4	3.5	3.7	3.9	4.0	3.9	3.8	3.6	3.7	3.6	3.5	3.7	3.9
152. Labor					1.6	2.0	2.2	2.3	2.2	2.1	2.0	2.1	2.0	2.0	2.1	2.1
153. Property					1.7	1.5	1.5	1.5	1.6	1.5	1.4	1.5	1.5	1.4	1.5	1.5
154. Taxes					0.2	0.1	0.1	0.1	0.1	0.1	0.1	0.1	0.1	0.1	0.1	0.1
155. Imports	0.0	0.0	0.0	0.0	0.1	0.1	0.1	0.1	0.1	0.1	0.1	0.1	0.1	0.1	0.1	0.1

Advertising																
156. Total	2.2	2.2	2.2	2.3	2.4	2.6	2.6	2.8	2.7	2.6	2.6	2.6	2.5	2.5	2.5	2.2
157. Labor					1.3	1.5	1.5	1.6	1.6	1.5	1.5	1.4	1.4	1.4	1.3	1.2
158. Property					1.0	0.9	1.0	0.9	0.9	1.0	1.0	1.0	1.0	1.0	1.0	0.9
159. Taxes					0.1	0.1	0.1	0.1	0.1	0.1	0.1	0.1	0.1	0.1	0.1	0.1
160. Imports	0.1	0.1	0.1	0.1	0.1	0.1	0.1	0.1	0.1	0.1	0.1	0.1	0.1	0.1	0.1	0.1
Legal, accounting, and bookkeeping																
161. Total	1.2	1.1	1.2	1.2	1.2	1.3	1.5	1.6	1.8	1.9	2.0	1.9	1.9	1.9	1.8	2.0
162. Labor					0.7	0.8	0.9	1.0	1.1	1.2	1.2	1.1	1.1	1.1	1.1	1.2
163. Property					0.5	0.5	0.5	0.5	0.6	0.6	0.7	0.7	0.7	0.7	0.7	0.7
164. Taxes					0.0	0.0	0.0	0.0	0.1	0.1	0.1	0.1	0.1	0.1	0.1	0.1
165. Imports	0.0	0.0	0.0	0.0	0.0	0.0	0.0	0.0	0.0	0.0	0.0	0.0	0.0	0.0	0.0	0.0

Source: USDA, Economic Research Service.

The foodservice industry led total food dollar value contributions (in nominal dollars) of the supply chain industry groups (table 3, line 31), followed by the food processing industry (table 3, line 11). However, value added to the food dollar by these two industry groups has been going in different directions. A comparison of the contribution of the two groups between the 1990s and the fi rst decade of the 2000s shows that food-services added about 8.3 cents in value to each dollar spent on food, an increase of over 30 percent, whereas food processors did not increase their value contribution. The largest percentage increase in value contribution among the industries was 74 percent from the energy industry group (table 3, line 36), followed by an increase of 38 percent from the legal/accounting/ bookkeeping industry group (table 3, line 51).

For at-home and food-away expenditures, the range of total returns to primary factors for salaries and output taxes over the study period was lower for the at-home food dollar than the food-away food dollar, whereas the range of total payments for imports and returns to capital was lower for total food-away food expenditures. For the food-away expenditures, the foodservice industry group dominated supply chain value added, accounting for about three-quarters of total value, and this industry group generated a higher than average share of its value added from salaries and retail sales taxes (table 3, lines 141-145). For at-home expenditures, the food processing industry accounted for close to a third of total food value over the study period (table 3, line 66), and the food retailing industry group accounted for about a quarter (table 3, line 81). Among supply chain industry groups, the changes in value contributions of the energy industry were the most pronounced, rising from 4.3 cents per dollar of at-home food expenditures in 1998 to 8.2 cents in 2008. About 90 percent of this increase was from higher energy imports and higher returns to capital.

Table 4 reports the value components of the U.S. real (infl ation-adjusted) expenditures of total, at-home, and food-away food dollars from 1993 to 2008. The table shows results for total domestic primary factor returns, total import primary factor returns, and the sum of those two factor returns, for all industries combined and each of the 10 supply chain industry groups, for a total of 99 categories. Table lines are numbered for reference.

A comparison of the nominal total industry import value added (table 3, line 5) and the corresponding real (infl ation-adjusted) series (table 4, line 3) shows a close correspondence over the period 1993 to 2000—both dollar series climbed from 4.7 cents to 6.0 cents. Between 2000 and 2006, however, the nominal series became volatile, dropping to 5.5 cents by 2002 and rising to

A Revised and Expanded Food Dollar Series   41

7 cents by 2006, whereas the real import value-added series remained about the same over the period. By 2008, the nominal series had reached 7.8 cents and the real series declined to 5.6 cents. Because the real import value-added industry series in table 4 is measured in constant year 2000 prices for domestic and imported commodities, the volatility of post-2000 nominal import values (table 3) in relation to the real import values (table 4) over this period refl ects volatility in the prices of imports relative to domestic commodities used by the food system. Notably, nominal import value added reached 9.7 cents of the at-home food dollar (table 3, line 60) in 2008, compared with a real (infl ation-adjusted) value of 7.1 cents (table 4, line 36). Imported energy, primarily crude oil and petroleum products, is behind this surge in import value. Energy is the only import commodity group to have an increase in real value added over the period 2000-08, even as the price of imported energy increased rapidly over this interval (table 4, line 24).

Between 1996 and 2005, the real (infl ation-adjusted) value added by food services gradually increased from 27 to 32 cents of each dollar spent on U.S.-produced food (table 4, line 19). Over this same interval, the price of food services remained roughly equal to their year 2000 level. In 2006, the price of food services declined, and the volume of these services purchased per dollar spent on food spiked 14 percent. In 2007-08, the average price of food services rose but remained below their year 2000 level, and the volume of purchases remained high, keeping nominal foodservice value added near its 2006 level.

The volume of retail trade services purchased in each real food dollar generally increased between 1993 and 2007, from 12.9 to 15.4 cents, but dropped off in 2008 to 13.3 cents of the total food dollar (table 4, line 16). Trade services per nominal food dollar remained in the 14- to 15-cent range over this interval, implying that the price of retail food services per unit of volume declined over the interval before spiking upward in 2008.

Figure 6 provides some context for these findings. Using annual data on the total number of supermarkets, full-service restaurants, and limited-service restaurants per capita in the United States for the years 2000-08, the figure charts these data as indexes representing the percentage of their respective year 2000 per capita establishment-count totals. For example, the chart indicates that there were roughly 10 percent fewer supermarkets per capita in the United States in 2008 than in 2000. Fewer food retail stores per capita would indicate that food retailing services per establishment had increased, suggesting that declines in the average price of food retailing services over the study period might have resulted from increasing economies of scale in the provision of the services.

Table 4. Cost components of the U.S. real food dollar, 1993 to 2008

Primary factor cost	1990s							2000s								
	93	94	95	96	97	98	99	2000	01	02	03	04	05	06	07	08
Total food dollar																
All industries																
1. Total	100.0	100.0	100.0	100.0	100.0	100.0	100.0	100.0	100.0	100.0	100.0	100.0	100.0	100.0	100.0	100.0
2. Val. added	95.3	95.4	95.1	94.9	94.6	94.4	94.2	94.0	94.1	94.1	94.0	94.0	93.8	94.0	93.6	94.4
3. Imports	4.7	4.6	4.9	5.1	5.4	5.6	5.8	6.0	5.9	5.9	6.0	6.0	6.2	6.0	6.4	5.6
Farm and agribusiness																
4. Total	13.2	12.0	12.7	13.1	12.8	12.7	12.7	12.4	12.2	11.3	11.6	11.2	10.8	10.0	10.7	10.8
5. Val. added	12.0	10.9	11.4	11.9	11.5	11.3	11.4	11.1	11.0	10.1	10.4	10.1	9.6	8.8	9.4	9.6
6. Imports	1.2	1.0	1.2	1.2	1.3	1.4	1.4	1.3	1.2	1.2	1.2	1.1	1.3	1.2	1.3	1.1
Food processing																
7. Total	23.2	24.6	23.9	23.9	24.1	23.5	22.5	21.7	21.1	20.9	20.4	19.7	20.3	18.6	19.2	19.2
8. Val. added	22.3	23.6	23.0	22.9	23.0	22.3	21.3	20.6	20.0	19.9	19.4	18.6	19.2	17.7	18.1	18.3
9. Imports	0.9	0.9	1.0	1.0	1.1	1.2	1.2	1.2	1.1	1.0	1.1	1.1	1.1	1.0	1.1	0.9
Packaging																
10. Total	4.9	5.0	5.1	5.0	4.8	4.9	4.8	4.7	4.7	4.8	4.6	4.4	4.4	4.1	4.0	3.8
11. Val. added	4.1	4.2	4.2	4.1	3.9	3.9	3.8	3.7	3.7	3.7	3.6	3.3	3.3	3.0	3.0	2.8
12. Imports	0.8	0.8	0.9	0.9	0.9	1.0	1.0	1.0	1.0	1.1	1.1	1.1	1.1	1.1	1.0	1.0
Transportation																
13. Total	4.9	5.0	4.8	4.8	4.7	4.7	4.4	4.1	4.0	4.5	4.0	4.1	4.2	4.2	4.0	3.9
14. Val. added	4.7	4.8	4.6	4.6	4.5	4.4	4.1	3.8	3.8	4.2	3.7	3.8	3.9	4.0	3.7	3.7
15. Imports	0.2	0.2	0.2	0.2	0.3	0.3	0.3	0.3	0.2	0.3	0.3	0.3	0.2	0.3	0.3	0.2
Retail trade																
16. Total	12.9	13.3	13.3	13.9	14.1	14.3	14.5	14.2	14.6	15.1	15.7	15.1	15.4	14.4	15.4	13.3
17. Val. added	12.7	13.1	13.1	13.7	13.8	14.1	14.2	13.9	14.3	14.8	15.4	14.8	15.1	14.2	15.1	13.1
18. Imports	0.2	0.2	0.2	0.2	0.3	0.3	0.3	0.3	0.3	0.3	0.3	0.3	0.3	0.2	0.3	0.2

Foodservice																
19. Total	28.0	28.0	28.1	27.1	27.5	27.7	28.3	29.4	29.9	30.0	30.5	32.4	31.6	36.0	33.5	35.7
20. Val. added	27.6	27.6	27.6	26.7	27.0	27.1	27.7	28.8	29.3	29.5	30.0	31.9	31.1	35.4	32.9	35.2
21. Imports	0.4	0.4	0.4	0.5	0.5	0.6	0.6	0.6	0.6	0.5	0.5	0.6	0.5	0.6	0.6	0.5
Energy																
22. Total	5.1	4.7	4.6	4.6	4.4	4.3	4.6	4.8	4.9	5.2	4.9	4.8	5.1	4.6	4.9	5.0
23. Val. added	4.3	3.9	3.8	3.8	3.5	3.6	3.8	3.8	3.8	3.8	3.6	3.4	3.7	3.1	3.3	3.6
24. Imports	0.8	0.8	0.8	0.8	0.8	0.7	0.8	1.0	1.1	1.4	1.3	1.4	1.5	1.5	1.5	1.4
Finance and Insurance																
25. Total	3.6	3.6	3.4	3.4	3.5	3.7	3.9	4.3	4.4	4.3	4.2	4.4	4.4	4.3	4.6	4.6
26. Val. added	3.5	3.5	3.3	3.3	3.4	3.6	3.8	4.2	4.2	4.1	4.1	4.3	4.2	4.1	4.4	4.5
27. Imports	0.1	0.1	0.1	0.1	0.1	0.1	0.1	0.2	0.1	0.1	0.1	0.1	0.1	0.1	0.1	0.2
Advertising																
28. Total	2.5	2.5	2.6	2.6	2.7	2.7	2.7	2.7	2.5	2.3	2.3	2.3	2.3	2.3	2.3	2.1
29. Val. added	2.4	2.4	2.5	2.6	2.6	2.6	2.6	2.6	2.4	2.3	2.2	2.3	2.2	2.2	2.2	2.0
30. Imports	0.1	0.1	0.1	0.1	0.1	0.1	0.1	0.1	0.1	0.1	0.1	0.1	0.1	0.1	0.1	0.1
Legal, accounting, and bookkeeping																
31. Total	1.6	1.5	1.5	1.5	1.5	1.5	1.5	1.6	1.7	1.7	1.7	1.6	1.6	1.6	1.5	1.6
32. Val. added	1.5	1.5	1.5	1.5	1.4	1.5	1.5	1.6	1.6	1.7	1.7	1.6	1.6	1.5	1.5	1.6
33. Imports	0.0	0.0	0.0	0.0	0.0	0.0	0.0	0.0	0.0	0.0	0.0	0.0	0.0	0.0	0.0	0.0

Table 4. (Continued)

Primary factor cost	1990s							2000s								
	93	94	95	96	97	98	99	2000	01	02	03	04	05	06	07	08
At-home food dollar																
All industries																
34. Total	100.0	100.0	100.0	100.0	100.0	100.0	100.0	100.0	100.0	100.0	100.0	100.0	100.0	100.0	100.0	100.0
35. Val. added	94.2	94.4	94.0	93.8	93.5	93.3	93.1	92.9	93.1	93.1	93.0	92.9	92.5	92.6	92.3	93.1
36. Imports	5.8	5.6	6.0	6.2	6.5	6.7	6.9	7.1	6.9	6.9	7.0	7.1	7.5	7.4	7.7	6.9
Farm and agribusiness																
37. Total	18.5	16.7	17.8	18.0	17.7	17.6	17.9	17.7	17.5	16.1	16.6	16.5	15.8	15.8	15.7	16.9
38. Val. added	16.8	15.2	16.1	16.3	15.9	15.7	16.0	15.9	15.8	14.5	14.9	14.9	14.0	14.0	13.8	15.2
39. Imports	1.7	1.4	1.7	1.7	1.7	1.9	1.9	1.8	1.7	1.6	1.6	1.6	1.8	1.8	1.9	1.7
Food processing																
40. Total	33.2	35.2	34.2	33.6	33.9	33.0	32.2	31.9	31.2	31.0	30.6	30.5	31.1	31.0	30.7	32.1
41. Val. added	31.9	33.9	32.8	32.2	32.4	31.4	30.5	30.2	29.6	29.5	29.0	28.9	29.4	29.4	29.0	30.7
42. Imports	1.3	1.3	1.3	1.4	1.5	1.7	1.7	1.7	1.6	1.5	1.6	1.7	1.7	1.6	1.7	1.4
Packaging																
43. Total	6.5	6.7	6.7	6.5	6.2	6.1	5.9	5.7	5.6	5.5	5.3	5.1	5.1	4.8	4.7	4.6
44. Val. added	5.4	5.6	5.5	5.3	5.1	4.9	4.7	4.5	4.3	4.3	4.1	3.8	3.8	3.5	3.5	3.4
45. Imports	1.0	1.1	1.2	1.2	1.2	1.2	1.2	1.3	1.2	1.2	1.2	1.2	1.3	1.2	1.2	1.2
Transportation																
46. Total	6.4	6.5	6.3	6.2	6.1	6.1	5.8	5.5	5.5	6.2	5.5	5.8	5.8	6.2	5.7	5.9
47. Val. added	6.2	6.2	6.1	5.9	5.8	5.7	5.4	5.1	5.1	5.8	5.1	5.4	5.5	5.9	5.3	5.5
48. Imports	0.2	0.2	0.3	0.3	0.4	0.4	0.4	0.4	0.3	0.4	0.4	0.4	0.3	0.4	0.5	0.3

Retail trade																
49. Total	20.1	20.4	20.6	21.1	21.6	22.7	23.5	23.9	25.3	26.5	27.8	27.8	27.7	28.1	28.7	25.6
50. Val. added	19.8	20.1	20.2	20.7	21.1	22.2	22.9	23.4	24.8	26.0	27.2	27.3	27.2	27.6	28.1	25.2
51. Imports	0.3	0.3	0.4	0.4	0.4	0.5	0.5	0.6	0.5	0.5	0.6	0.5	0.5	0.5	0.6	0.4
Foodservice																
52. Total	0.9	0.9	0.9	0.9	0.9	0.9	0.9	0.9	0.9	0.9	0.9	0.9	0.8	0.8	0.8	0.8
53. Val. added	0.9	0.9	0.9	0.9	0.9	0.9	0.9	0.9	0.9	0.9	0.9	0.8	0.8	0.8	0.8	0.8
54. Imports	0.0	0.0	0.0	0.0	0.0	0.0	0.0	0.0	0.0	0.0	0.0	0.0	0.0	0.0	0.0	0.0
Energy																
55. Total	6.6	6.0	5.9	5.8	5.6	5.3	5.5	5.5	5.4	5.4	5.2	5.1	5.4	4.9	5.2	5.5
56. Val. added	5.5	5.0	4.8	4.8	4.5	4.4	4.4	4.3	4.1	3.9	3.7	3.6	3.8	3.2	3.4	3.8
57. Imports	1.1	1.1	1.1	1.1	1.1	0.9	1.1	1.2	1.3	1.5	1.5	1.6	1.7	1.7	1.8	1.7
Finance and insurance																
58. Total	3.8	3.7	3.6	3.7	3.8	4.1	4.2	4.6	4.6	4.4	4.2	4.4	4.5	4.5	4.7	5.0
59. Val. added	3.8	3.7	3.6	3.7	3.8	4.0	4.2	4.5	4.5	4.3	4.2	4.4	4.4	4.4	4.7	4.9
60. Imports	0.0	0.0	0.0	0.0	0.0	0.1	0.1	0.1	0.1	0.1	0.1	0.1	0.1	0.1	0.1	0.1
Advertising																
61. Total	2.2	2.2	2.3	2.4	2.4	2.5	2.5	2.7	2.5	2.4	2.4	2.4	2.4	2.4	2.4	2.2
62. Val. added	2.2	2.2	2.3	2.3	2.4	2.5	2.5	2.6	2.4	2.4	2.3	2.4	2.3	2.3	2.3	2.1
63. Imports	0.1	0.1	0.1	0.1	0.1	0.1	0.1	0.1	0.1	0.1	0.1	0.1	0.1	0.1	0.1	0.1
Legal, accounting, and bookkeeping																
64. Total	1.8	1.7	1.7	1.7	1.7	1.7	1.6	1.6	1.6	1.5	1.5	1.5	1.5	1.4	1.4	1.4
65. Val. added	1.8	1.7	1.7	1.7	1.7	1.6	1.6	1.6	1.6	1.5	1.5	1.4	1.4	1.4	1.3	1.4
66. Imports	0.0	0.0	0.0	0.0	0.0	0.0	0.0	0.0	0.0	0.0	0.0	0.0	0.0	0.0	0.0	0.0

Table 4. (Continued)

Primary factor cost	1990s							2000s								
	93	94	95	96	97	98	99	2000	01	02	03	04	05	06	07	08
Away-from-home food dollar																
All industries																
67. Total	100.0	100.0	100.0	100.0	100.0	100.0	100.0	100.0	100.0	100.0	100.0	100.0	100.0	100.0	100.0	100.0
68. Val. added	96.6	96.7	96.4	96.3	96.1	95.8	95.7	95.5	95.7	95.7	95.7	95.7	95.6	95.7	95.4	96.1
69. Imports	3.4	3.3	3.6	3.7	3.9	4.2	4.3	4.5	4.3	4.3	4.3	4.3	4.4	4.3	4.6	3.9
Farm and Aagribusiness																
70. Total	4.9	4.3	4.4	4.7	4.7	4.8	4.8	4.8	5.0	4.7	5.2	4.8	4.5	3.7	4.9	4.0
71. Val. added	4.4	3.8	3.9	4.2	4.1	4.2	4.2	4.2	4.4	4.1	4.5	4.3	3.9	3.2	4.2	3.5
72. Imports	0.5	0.4	0.5	0.5	0.5	0.6	0.6	0.6	0.6	0.6	0.6	0.6	0.6	0.5	0.7	0.5
Food processing																
73. Total	6.0	5.9	6.0	6.1	6.4	7.4	7.3	7.1	7.4	7.8	7.6	6.9	7.1	5.7	5.8	5.0
74. Val. added	5.6	5.6	5.6	5.7	6.0	7.0	6.8	6.7	7.0	7.4	7.1	6.5	6.7	5.3	5.4	4.7
75. Imports	0.3	0.3	0.3	0.4	0.4	0.5	0.5	0.5	0.5	0.5	0.5	0.4	0.5	0.4	0.4	0.3
Packaging																
76. Total	3.4	3.5	3.6	3.5	3.4	3.5	3.5	3.4	3.3	3.3	3.2	3.1	3.1	2.9	2.8	2.6
77. Val. added	2.9	2.9	2.9	2.8	2.8	2.8	2.7	2.6	2.6	2.6	2.4	2.3	2.3	2.1	2.1	1.9
78. Imports	0.6	0.6	0.6	0.6	0.7	0.7	0.7	0.8	0.7	0.8	0.7	0.8	0.8	0.8	0.7	0.7
Transportation																
79. Total	2.6	2.6	2.5	2.5	2.4	2.5	2.3	2.1	2.0	2.2	2.1	2.1	2.1	2.1	1.9	1.8
80. Val. added	2.5	2.5	2.4	2.3	2.3	2.3	2.1	1.9	1.9	2.0	1.9	1.9	1.9	1.9	1.8	1.7
81. Imports	0.1	0.1	0.1	0.2	0.2	0.2	0.2	0.2	0.1	0.2	0.2	0.2	0.1	0.1	0.2	0.1

Category																
Retail trade																
82. Total	0.5	0.6	0.6	0.6	0.6	0.6	0.6	0.5	0.4	0.5	0.5	0.4	0.5	0.4	0.5	0.4
83. Val. added	0.5	0.6	0.6	0.6	0.6	0.6	0.6	0.5	0.4	0.5	0.5	0.4	0.5	0.4	0.5	0.4
84. Imports	0.0	0.0	0.0	0.0	0.0	0.0	0.0	0.0	0.0	0.0	0.0	0.0	0.0	0.0	0.0	0.0
Foodservice																
85. Total	70.5	71.6	71.5	71.0	71.0	69.5	69.7	69.9	69.9	69.8	70.1	71.2	71.1	74.0	72.4	74.6
86. Val. added	69.5	70.5	70.3	69.7	69.6	68.0	68.2	68.3	68.5	68.5	68.8	69.9	69.8	72.7	71.1	73.4
87. Imports	1.1	1.1	1.2	1.3	1.4	1.5	1.5	1.6	1.4	1.3	1.3	1.3	1.3	1.3	1.4	1.2
Energy																
88. Total	4.4	4.1	4.0	4.0	3.8	3.8	3.8	3.8	3.8	3.9	3.7	3.7	3.8	3.5	3.7	3.8
89. Val. added	3.8	3.5	3.4	3.4	3.2	3.2	3.2	3.1	3.0	3.0	2.8	2.7	2.9	2.5	2.7	2.9
90. Imports	0.6	0.6	0.6	0.6	0.6	0.6	0.6	0.7	0.8	0.9	0.9	0.9	1.0	1.0	1.1	0.9
Finance and insurance																
91. Total	3.3	3.3	3.2	3.3	3.4	3.6	3.7	4.0	4.0	3.8	3.7	3.9	3.9	3.9	4.1	4.2
92. Val. added	3.3	3.2	3.1	3.3	3.3	3.5	3.7	3.9	3.9	3.8	3.7	3.8	3.9	3.8	4.1	4.1
93. Imports	0.0	0.0	0.0	0.0	0.0	0.0	0.1	0.1	0.1	0.1	0.1	0.1	0.1	0.1	0.1	0.1
Advertising																
94. Total	2.4	2.4	2.5	2.5	2.6	2.6	2.7	2.8	2.5	2.5	2.4	2.5	2.4	2.4	2.4	2.2
95. Val. added	2.4	2.3	2.4	2.5	2.5	2.6	2.6	2.7	2.5	2.4	2.4	2.4	2.4	2.4	2.3	2.2
96. Imports	0.1	0.1	0.1	0.1	0.1	0.1	0.1	0.1	0.1	0.1	0.1	0.1	0.1	0.1	0.1	0.1
Legal, accounting, and bookkeeping																
97. Total	1.8	1.7	1.7	1.7	1.7	1.7	1.6	1.6	1.6	1.5	1.5	1.5	1.5	1.4	1.4	1.4
98. Val. added	1.8	1.7	1.7	1.7	1.7	1.6	1.6	1.6	1.6	1.5	1.5	1.5	1.5	1.4	1.3	1.4
99. Imports	0.0	0.0	0.0	0.0	0.0	0.0	0.0	0.0	0.0	0.0	0.0	0.0	0.0	0.0	0.0	0.0

Source: USDA, Economic Research Service.

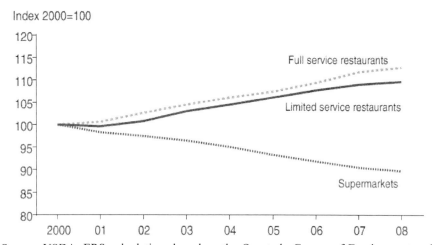

Source: USDA, ERS calculations based on the Quarterly Census of Employment and Wages, Bureau of Labor Statistics, U.S. Department of Labor, and Annual Population Estimates, Census Bureau, U.S. Department of Commerce.

Figure 6. Number of establishments per capita, 2000-08.

A comparison of table 3, line 141 and table 4, line 85 shows that the nominal foodservice value added per food-away dollar ran slightly higher than the real (inflation-adjusted) volume shares between 2000 and 2004, but beginning in 2005 the nominal foodservice share fell below real share value. Since the nominal foodservice share also gained about 4 cents of the overall food-away dollar between 2000 and 2004, these results indicate that the average cost of food services increased relative both to its year 2000 price and to overall food prices between 2000 and 2004. From 2005 to 2008, average foodservice prices fell below their year 2000 level, but the nominal foodservice share of the food-away dollar continued to increase due to growth in the volume of foodservices in the food-away dollar. Returning to figure 6, note that from 2000 forward, the rate of increase in the number of full-service restaurants per capita outpaced the increase in limited-service restaurants. This suggests that U.S. food consumers were purchasing more services per meal over time, which could explain at least some of both the price and volume changes in foodservices over the period.

THE FOOD DOLLAR: LOOKING AHEAD

Historical observations of relationships among food dollar expenditures, the component values of those expenditures, and both producer and consumer food prices can inform our understanding of the implications of USDA food market forecasts for the food dollar.

Several determinants of change in U.S. food markets between 1993 and 2008 are especially notable. The increasing demand for food services by U.S. households, followed by an increased demand for higher priced services (e.g., full- versus limited-service restaurants), was instrumental in driving down the farm share of the food dollar over a portion of the study period. Even in periods where the relative price of farm commodities appeared to be gaining on retail food prices, increasing purchases of food services—essentially, paying others to prepare meals and clean up afterwards—drove the farm share and the farm and agribusiness value-added shares down. A weakening of the U.S. dollar relative to other international currencies drove up the prices of imported ingredients for the food system. Yet, while the real value of most imports remained a roughly constant share of the real food dollar over the study period (table 4), the real value of imported energy increased by over 50 percent, and this—coupled with the sharp increase in energy prices starting in 2003—pushed the import share of the nominal food dollar up by more than 300 percent overall (1993 to 2008), to as high as 8.2 cents for the 2008 at-home food dollar. Although the food retailing nominal value-added share was largely unchanged over the study period, an examination of the constant price value share of food retailing shows that food retailing services substantially increased over the interval, but these volume increases were offset by equally substantial price decreases for these services.

Energy value shares can be infl uenced by energy prices, but can also be affected by the types of food products households purchase, as well as by labor market conditions. Between 1997 and 2007, the U.S. food system increased its energy use, and both tight labor markets and consumer demand for more food processing services were found to loom large in these increases (Canning et al., 2010). Rapid consolidation of the food retailing industry and economies of scale appear to be behind the decline in price and increase in volume of food retailing services over the study period. Continuing consolidation in the food retailing industry may produce more scale economies, but these could be offset by increased incentives for price collusion.

A few observations stand out regarding the new estimation methodology described in this report:

- A lower farm share measure is obtained in the new food dollar series than the measure under the old marketing bill, largely accounted for by the deduction of farm-to-farm transactions, as specifi ed in equation 2.
- Comparing real (infl ation-adjusted) estimates of the farm share with nominal (current-value) estimates allows us to see to what extent changes in the farm share over time are attributable to changes in farm-level prices relative to other prices in the economy, and to what extent the farm share changes are attributable to changes in the quantity of services and materials provided by the food marketing system.
- Separately comparing changes in the farm share for food away from home and food at home allows us to see to what extent changes in the farm share over time are attributable to the increase in food consumed away from home.
- Estimates of value added by individual sectors of the food supply chain provide a new, more comprehensive way of looking at where the consumer's food dollar goes.
- Dividing the food dollar by sector, and then separately by factor, avoids the potentially confusing divisions of the marketing bill method, which combined the two.

Theoffi cial USDA annual outlook projections are based on clear and defensible assumptions about key market outcomes (U.S. Department of Agriculture, 2010). Presented as a point of departure for discussions of alternative farm and food industry outcomes, the current long-term report includes projections out to 2019 for U.S. farm-level food commodity prices, consumer food prices for both at-home and food-away expenditures, and total at-home and food-away food expenditures (fi gure 7). Farm commodity prices are projected to initially lose ground for a few years on retail food prices and then to keep pace through most of the next decade. Food-away expenditures are forecast to outpace at-home expenditures over this period. Both sets of projections point to a falling farm share measure.

Year-to-year percent change in price indexes

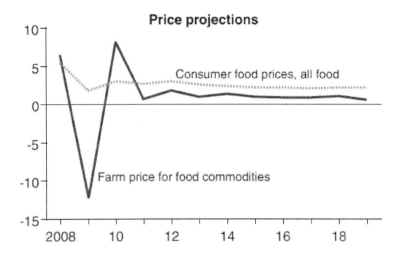

Annual U.S. food expenditures - $ billion

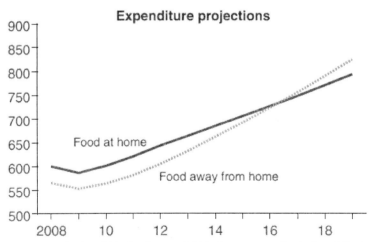

Source: USDA Agricultural Projections to 2019, Interagency Agricultural Projections Committee/USDA, 2010.

Figure 7. Long-term U.S. price and expenditure projections, 2008-19.

REFERENCES

Canan, B., and R.W. Cotterill. 2006. "Strategic Pricing in a Differentiated Product Oligopoly Model: Fluid Milk in Boston," *Agricultural Economics* 35:27-33.

Canning, P., A. Charles, S. Huang, K.R. Polenske, and A. Waters. 2010. *Energy Use in the U.S. Food System*. U.S. Department of Agriculture, Economic Research Service, Economic Research Report No. 94 (March). http://www.ers.usda.gov/Publications/err94/err94.pdf

Elitzak, H. 1999. *Food Cost Review, 1950-97*. U.S. Department of Agriculture, Economic Research Service, Agricultural Economic Report No. 780 (June):19-23. http://www.ers.usda.gov/Publications/AER780/

Elitzak, H. 2004. "Calculating the Food Marketing Bill," *Amber Waves*, Vol. 2, No. 1 (February): p. 43. U.S. Department of Agriculture, Economic Research Service. http://www.ers.usda.gov/AmberWaves/February04/

Gale, H.F. 1967. *The Farm Food Marketing Bill and its Components*. U.S. Department of Agriculture, Economic Research Service, Agricultural Economics Report No. 105 (January).

Gardner, B. 1975. "The Farm-Retail Price Spread in a Competitive Food Industry," *American Journal of Agricultural Economics* 57(3):399-409.

Golan, Amos, George Judge, and Sherman Robinson. 1994. "Recovering Information From Incomplete or Partial Multisectoral Economic Data," *The Review of Economics and Statistics* LXXVI(3):541-49.

Harp, H. 1987. Major Statistical Series of the U.S. Department of Agriculture, Vol. 4: *Agricultural Marketing Costs and Charges*. Agricultural Handbook No. 671.

Hirst, E. 1974. "Food-Related Energy Requirements," *Science* 184(April 12):134-38.

Kuchler, F., and O. Burt. 1990. "Revisions in the Farmland Value Series." Special Article for *Agricultural Resources: Agricultural Land Values and Markets Situation and Outlook Report*, AR-18 (June), U.S. Department of Agriculture, Economic Research Service, pp. 32-35.

Leontief, W. 1986. *Input-Output Economics*. 2nd ed. New York: Oxford University Press.

McDougall, R.A. 1999. "Entropy Theory and RAS are Friends." Paper presented at the 5th Conference of Global Economic Analysis, Copenhagen, Denmark.

Schluter, G., C. Lee, and M. LeBlanc. 1998. "The Weakening Relationships between Farm and Food Prices," *American Journal of Agricultural Economics* 80(5):1134-38.

U.S. Department of Agriculture, Offi ce of the Chief Economist. 2010. *USDA Agricultural Projections to 2019*, OCE-2010-1 (February). http://www.ers.usda.gov/publications/oce101/

U.S. Department of Commerce, Bureau of Economic Analysis. (various years). *Benchmark Input-Output Accounts*, available at: www.bea.gov/industry/index.htm#benchmark_io

U.S. Department of Commerce, Bureau of Economic Analysis. (various years). *Annual Input-Output Accounts*, available at: http://www.bea.gov/industry/index.htm#annual

U.S. Department of Commerce, U.S. Census Bureau. *Annual Population Estimates 2000 to 2008*, www.census.gov/popest/states

U.S. Department of Labor, Bureau of Labor Statistics. (various years). *Inter-industry Relationships (Input/Output matrix)*, available at: www.bls.gov/emp/ep_data_input_output_matrix.htm

U.S. Department of Labor, Bureau of Labor Statistics (BLS). (various years). *Quarterly Census of Employment and Wages*, www.bls.gov/cew/

APPENDIX: DATA SOURCES AND MATHEMATICAL DERIVATION OF THE FOOD DOLLAR SERIES

A direct measure of the *food dollar* and the *farm share*, using input-output (IO) analysis, is stated concisely in equation 3 of this report, and measurement of value-added food dollar components is stated in equation 5. Although these measures are a conventional application of IO analysis, substantial preliminary data processing is needed to compile the source data into the required structure for implementing this analysis.

Data Sources

The principal source data for this report are the annual U.S. input-output tables published by the U.S. Department of Labor's Bureau of Labor Statistics (BLS). The BLS IO data series is released at 2-year intervals, with each release adding two annual data tables to the time series. The most recent release, in December 2009, provides annual tables for 1993 to 2008 (www. bls.gov/emp/). Ancillary source data on industry value added by primary production factors are obtained from the annual industry Make and Use tables (before redefi nitions) published each year by the U.S. Department of Commerce's Bureau of Economic Analysis (BEA). A mid-2010 release by BEA includes annual Make and Use tables for 1998 to 2008 (www.bea.gov/ industry/). The IO-based food dollar series in the present report are benchmarked to give more detailed estimates, using 1997 and 2002 benchmark IO Make and Use tables, also published by BEA. These benchmark accounts are released in 5-year intervals, with a 5-year lag between enumeration and release. For example, final release of the 2002 benchmark account was in mid-2008.

Additional source data are required to address a problem of aggregation bias in the available IO accounts. Although wholesale and retail trade margins are directly measured from primary data sources in the IO accounts, the composite wholesale and retail industries that facilitate these margin services are assumed to employ identical technologies for all users. As discussed in Hirst (1974), this convention produces an aggregation bias for the energy use by trade industry establishments serving the food system. The Economic Census (www.census. gov/econ/) and Annual Survey of Retail Trade (www.census.gov/retail/) provide data for electricity and natural gas utility expenses per dollar of trade margin revenues for several 4-digit wholesale and retail North American Industry Classifi cation System (NAICS) industry groups (www.census.gov/ eos/www/naics/). These data are integrated into the IO accounts to address the identifi ed aggregation bias (discussed in the next section).

To implement the IO estimation procedure for the new food dollar series, all five data sources discussed above must be fully integrated, and a concordance matrix was developed to map the relationships among the industry/ commodity aggregations of the five accounts. To facilitate a clean concordance across these NAICS-based data systems, some aggregation of the BLS industry/commodity groups was required. As a starting point for the data compilation procedures, the 202-order BLS IO tables are reduced to a 184-order aggregated annual IO table, the 400-plus order 1997 and 2002 benchmark BEA tables are both reduced to the same 390-order aggregated benchmark tables, and the 92-order annual BEA JO tables are reduced to a 90-order aggregated JO table.

Disaggregation and Decomposition of IO Accounts

The basic building blocks of input-output analysis are the "make," "use," "value added," and "fi nal demand" tables, denoted by the matrices M and U and the vectors v and f, respectively.[13] For a given year, the make table itemizes the value for production of each commodity $c \in C$ by each industry $i \in I$. The use table itemizes annual outlays of each industry $i \in I$ for purchases of each commodity $c \in C$. The value-added table itemizes annual outlays of each industry $i \in I$ for total payments to primary factor owners $p \in P$. The final demand table itemizes total final market expenditures on each commodity $c \in C$. These four building block tables are published biannually by BLS.

Additionally, BLS publishes a supplemental fi nal demand table that provides a detailed breakout of fi nal demand expenditures. This data, along with the benchmark and annual make and use tables published by BEA, facilitates the computation of various share and binary vectors used extensively in this appendix to disaggregate and decompose the BLS JO accounts. The share vectors are denoted s_id, where id identifi es the share metric that is declared as needed. The binary vectors are denoted b_id, where id is declared as needed to identify a target commodity group. Binary vectors have unit values positioned in the appropriate sequence of a vector to identify target groups and have null values elsewhere.

Procedures for compiling the BLS building block tables to carry out the analysis in this report involve several steps:

For each annual table from 1993 to 2008, convert the commodity by industry use table into a commodity-by-commodity interindustry transaction table (Z):

$$Z_{C,C} = U_{C,I} \cdot (\overline{\{M_{I,C} \cdot i_C\}}^{-1} \cdot M_{I,C})$$

(A1)

The 'i' is a unit vector that, when post- (pre-) multiplied into a matrix, produces a vector of the matrix row (column) sums. Equation A1 consolidates the production of each commodity $c \in C$ by the different industries producing the commodity into a single composite industry, such that there is a one-to-one mapping of industries to commodities.

For each annual table from 1993 to 2008, convert the primary factor by the industry value-added table into an import-inclusive primary factor by commodity value-added table, and update the fi nal demand table accordingly.

Let s_m denote the vector representing import shares of available product for each commodity $c \in C$, as derived from the BLS supplemental fi nal demand table.[14] Then the import-inclusive commodity value-added table is as follows:

$$v\_m_{C,P} = \left[\left(v'_I \cdot \left(\{ \overline{M_{I,C} \cdot i_C} \}^{-1} \cdot M_{I,C} \right) \right)' \middle| \overline{s\_m_C} \cdot f_C \right]$$

(A2)

This change requires that final demand also be updated:

$$y_C = \overline{(i_C - S\_m_C)} \cdot f_C.$$

(A3)

To verify that the procedures are properly carried out, the following relationships must hold among the three expressions:

$$Z \cdot i + y = Z' \cdot i + v_m \cdot i$$

(A4)

Decouple the electricity and natural gas utility outlays of wholesale and retail service industries.

According to the 2002 Economic Census, food retailers spent an average of 3.9 cents on electric utilities for every dollar in retail margin revenues, whereas the average of this expenditure across all retail establishments was 1.3 cents. Available data facilitates the disaggregation of retail services into 11 distinct users' groups, requiring a wide range of electric utility purchase coeffi cients. Similar scenarios exist for the wholesale industry and for natural gas utilities in both industries. To capture this wide variability, wholesale and retail industry utility outlays are decoupled and consolidated into a new 'trade utilities' (*tu*) industry, as follows:

A Revised and Expanded Food Dollar Series 57

$$Z_{C,tu}^{new} = \overline{b\_ge_C} \cdot (Z_{C,wt} + Z_{C,rt}) \,, \tag{A5}$$

where b_ge is a binary vector identifying the natural gas and electric utility rows, tu denotes the position of the new trade utility industry column, and wt and rt denote the positions of the wholesale and retail trade industry columns, respectively. No value-added outlays are attributed to the new trade utility industry:

$$v\_m_{tu,P}^{new} = 0 \tag{A6}$$

The original wholesale and retail outlay columns are updated accordingly:

$$Z_{C,wt}^{new} = \overline{(i_c - b\_ge_C)} \cdot Z_{C,wt} \cdot \,, \tag{A7}$$

$$Z_{C,rt}^{new} = \overline{(i_c - b\_ge_C)} \cdot Z_{C,rt} \cdot \tag{A8}$$

Trade industry outlays for primary factor value added are unchanged.

Sales of the new trade utility commodity are measured based on the share of total outlays going to electricity and natural gas utilities for wholesale establishments selling each commodity $c \in C$ (s_we_C, s_wg_C), the share of total outlays going to electricity and natural gas utilities for retail establishments selling each commodity $c \in C$ (s_re_C, s_rg_C),[15] and the trade margin matrices summarizing the wholesale and retail margins added to fob[16] values reported in the interindustry transaction matrix (Z_wt, Z_rt):

$$Z_{tu,C}^{new} = (S\_we_C + S\_wg_C)' \cdot Z\_wt_{C,C} + (s\_re_C + s\_rg_C)' \cdot Z_{rt_{C,C}} \tag{A9}$$

No trade or transportation margin costs are attributed to the utility outlays, so if we denote tt the set of all trade and transportation margin industries (and the set of row/column positions of these industries in the transaction matrix), we have:

$$Z\_tt_{C,tu}^{new} = 0$$

(A10)

Final market sales of the new trade utility commodity are similarly computed:

$$y_{tu} = (S\_we_C + S\_wg_C)' \cdot y\_wt_C + (S\_re_C + S\_rg_C)' \cdot y_{rt_C}$$

(A11)

The appropriate portions of the proceeds in the new trade utility sales row are deducted from the wholesale and retail sales rows in the updated transaction matrix and final demand vector. No trade or transportation margin costs are attributed to the utility outlay, so we have:

$$y\_tt_{tu} = 0$$

(A12)

Decouple the foodservice margins from food-away expenditures.

In the IO accounts, food-at-home expenditures are separately recorded as fob food commodity purchases, along with the trade margins associated with these purchases. Conversely, nearly all food-away purchases are recorded as expenditures on foodservices, where the food commodities are indirectly purchased as part of the service. To facilitate calculations of food dollar expenditures, the foodservices (*fs*) are decoupled from the food-away (*fa*) commodity purchases:

$$Z_{C,fa}^{new} = \overline{b\_fa_C} \cdot Z_{C,fs},$$

(A13)

$$Z_{C,fs}^{new} = Z_{C,fs} - Z_{C,fa}^{new}.$$

(A14)

No value-added outlays are attributed to the new food-away industry:

$$v\_m_{fa,P}^{new} = 0,$$

(A15)

A Revised and Expanded Food Dollar Series

$$v\_m_{fs,P}^{new} = v\_m_{fs,P} \ .$$

(A16)

All intermediate and final market sales of the new food-away and foodservice commodities are fixed shares of the old consolidated foodservice commodity sales:

$$Z_{fa,C}^{new} = Z_{fs,C} \cdot s\_fa, \ ,$$

(A17)

$$Z_{fs,C}^{new} = Z_{fs,C} \cdot (1 - S\_fa \ ),$$

(A18)

where s_fa is computed as:

$$s\_fa = (i_C' \cdot Z_{C,fa}^{new}) \cdot \{i_C' \cdot Z_{C,fs} + v\_m_{fs,P} \cdot i_P\}^{-1} \ .$$

(A19)

The new trade and transport margin matrices are computed as:

$$Z\_tt_{C,fa}^{new} = \overline{b\_fa}_C \cdot Z\_tt_{C,fs} \cdot s\_fa \ ,$$

(A20)

$$Z\_tt_{C,fs}^{new} = Z_{tt C,fs} - Z\_tt_{C,fa}^{new} \ .$$

(A21)

Itemize total industry value-added outlays into primary factor payments to labor, other industry assets, and taxes on industry output.
Industry value-added tables of the BLS annual accounts are reported as total outlays to all primary factor owners. Both the BEA benchmark and annual accounts report primary factor payments to labor, other industry assets, and taxes on industry output. Because the 1997 and 2002 benchmark industry accounts map into the annual BLS accounts and are also used to benchmark the annual food dollar series, 1997 and 2002 primary factor payment shares calculated from the benchmark tables are applied to the BLS data for those years. For the years 1998 to 2008 (excluding 2002), primary factor payment

shares calculated from the nearest benchmark year tables are applied to the annual BLS data to formulate initial estimates, $v^0_{C,p}$. A maximum entropy mathematical programming model (Golan et. al., 1994) is employed to reconcile BEA annual data with more aggregated commodity group coverage (c^+ E C^+) but more detailed factor payment coverage ($p \in P$), and with BLS annual data having more detailed commodity groups ($c \in C$) but more aggregated factor payment data (P).

Minimize,

$$\sum_{c \in C} \sum_{p \in P} (v^0_{c,p})^{-1} \cdot Ln\left(\frac{\tilde{v}_{c,p}}{v^0_{c,p}}\right),$$

(A22)

Subject to,

$$\sum_{c \in C^+} \tilde{v}_{c,p} = v_{c+,p} \quad \forall \, c^+ \in C^+, p \in P,$$

(A23)

$$\sum_{p \in P} \tilde{v}_{c,p} = v_{c,p} \quad \forall \, c \in C.$$

(A24)

A solution to the model in equations A22 to A24 replicates both the annual BLS total factor payment values and the annual BEA detailed factor payment share values, while producing the minimum percentage difference between annual factor payment estimates and the nearest corresponding benchmark factor payment values. Since each entropy distance measure in the objective function (equation A22) is weighted by the reciprocal of its preliminary estimate, the maximum entropy solution can be solved iteratively using a RAS-type algorithm (McDougall, 1999). This is benefi cial, since there are occasional negative primary factor value-added outlays that are handled more routinely using RAS.[17]

A Revised and Expanded Food Dollar Series

Identify and measure the proportion of each fi nal demand commodity outlay that meets the criteria for food dollar or food and beverage dollar expenditure.

The BLS detailed final demand table identifi es food-at-home and food-away personal consumption expenditures and their accompanying margin costs, denoted *fh_pce* and *fa_pce*. Some food commodity outlays and the margin expenditures associated with them are not of farm origin, including salt, certain food products of the chemical industry, and water products sold by food retailers. All remaining food commodity expenditures are part of the food dollar bundle (*fd*). Dividing the total final demand into the food dollar produces the food dollar share value (including imports) of total fi nal demand:

$$S\_fd_c = \{\overline{y}_c\}^{-1} \cdot (\overline{b\_fd_c} \cdot (fa\_pce_C + fh\_pce_c)) \tag{A25}$$

If x_C is a vector reporting total domestic availability (domestic production plus imports) of each commodity $c \in C$, then the direct requirement matrix ($A_{C,C}$) summarizes outlays of each domestic industry $i \in I$ for each commodity $c \in C$ per unit of available industry product:

$$A_{C,C} = Z_{C,C}^{new} \cdot \{\overline{x}_C\}^{-1} \tag{A26}$$

The total requirement matrix, also called the Leontief matrix, is obtained as follows:

$$L_{C,C} = \{\overline{i}_C - A_{C,C}\}^{-1} \tag{A27}$$

From equations A27 and A25, total farm industry group (a) sales (including those from imports) required to accommodate total import-inclusive food dollar expenditures can be obtained as follows:

62 Patrick Canning

$$x_a^{fd} = L_{a,C} \cdot \underbrace{\overline{S\_fd}_C \cdot y_C}_{=y_C^{fd}}$$

(A28)

To avoid double counting of sales within the farm industry, all payments to a farm industry that are passed on and subsequently go directly or indirectly to another farm industry are netted out:[18]

$$x_a^{net} = (\bar{i}_a - A_{a,a} - \hat{A}_{a,a}) \cdot x_a^{fd} = x_a^{fd} - (\underbrace{A_{a,a}}_{\substack{farm-to-farm \\ direct}} + \underbrace{\hat{A}_{a,a}}_{\substack{farm-to-farm \\ indirect}}) \cdot x_a^{fd} ,$$

(A29)

where,

$$\hat{A}_{a,a} = (\sum\nolimits_{k=0}^{k*} A_{a,C} \cdot A_{C,C}^k) \cdot A_{C,a} .$$

(A30)

In equation A30, A^k refers to the matrix product (e.g., $A^2 = A \cdot A$) and $k*$ represents the exponent where the matrix product $Ak* \approx 0$.

Supply Chain Analysis to Measure
Value Components of the Food Dollar

A matrix reduction procedure in IO analysis (Leontief, 1986, chapter 3) to facilitate supply chain studies produces a clean decomposition of industry cost contributions by supply chain categories.

To demonstrate, organize industries/commodities into supply chain (*sc*) and non-chain (*nc*) groups respectively. The derivation for import-inclusive gross farm industry output in equation A28 is restated here, using the new supply chain partition:

$$x_{sc}^{fd} = L_{sc,sc} \cdot y_{sc}^{fd} + L_{sc,nc} \cdot y_{nc}^{fd}$$

(A31)

A Revised and Expanded Food Dollar Series 63

Equation A31 states that total outlays of supply chain industries refl ect their total requirements to meet final sales for the supply chain and the 'non-chain' commodities. This can be restated as a reduced supply chain input-output system by multiplying both sides through by the inverse of $L_{sc,sc}$ and rear-ranging terms:

$$\underbrace{y_{sc}^{fd} + L_{sc,sc}^{-1} \cdot L_{sc,nc} \cdot y_{nc}^{fd}}_{=y_{sc}^{*}} = L_{sc,sc}^{-1} \cdot x_{sc}^{fd}$$

(A32)

The innovation of this matrix reduction is the second inversion of the supply chain quadrant in the original total requirement matrix, $L_{sc,sc}$. It turns out that the second inversion of this submatrix produces the reduced structural supply chain matrix (see proof in Leontief, 1986):

$$L_{sc,sc}^{-1} = I_{sc} - \underbrace{[A_{sc,sc} + A_{sc,nc} \cdot (\bar{i}_{nc} - A_{nc,nc})^{-1} \cdot A_{nc,sc}]}_{=A_{sc,sc}^{*}}$$

(A33)

The reduced structural matrix $A*_{sc,sc}$ describes both the direct requirements of the supply chain interindustry purchases per unit of output and the supply chain industry requirements of the non-chain industries. Combining equations A32 and A33 and rearranging terms produces the supply chain input-output model:

$$x_{sc}^{fd} = (\bar{i}_{sc} - A_{sc,sc}^{*})^{-1} \cdot y_{sc}^{*}$$

(A34)

To use this model for conducting supply chain value-added analysis, the reduced supply chain primary factor value-added coeffi cient vector must also be obtained, and this is done using the same procedures applied to estimates of the reduced structural matrix:

$$v\_m^{*}_{sc,P} = v\_m^{*}_{sc,P} + [L'_{sc,sc}]^{-1} \cdot L'_{nc,sc} \cdot v\_m_{nc,P}$$

(A35)

In A35, total primary factor value added of non-chain industries per unit of output for each supply chain industry is added to the supply chain industries' direct value-added requirements. The resulting value-added coeffi cient vector captures the combined value contributions of each supply chain industry and the collective value added from all the subcontracted services. Supply chain analysis proceeds as follows:

$$i' \cdot y^{*}_{sc} = i'_{sc} \cdot (\overline{x}^{fd}_{sc} \cdot v\_m^{*}_{sc,P}) \cdot i_{P}$$

(A36)

Unlike conventional JO cost analysis, the vector within the brackets in equation A36 attributes total costs of final market expenditures to a short list of supply chain industries. To the extent that non-chain industries contribute to market costs, these costs are attributed to supply chain industries by a precise measure of the 'subcontracted' non-chain industry services provided to each stage of the analyzed commodity supply chains.

ACKNOWLEDGMENTS

The author benefi ted from reviews by Ken Hansen of USDA's Economic Research Service, Howard Leathers from the University of Maryland, and Nicolas Rockler from the Massachusetts Institute of Technology, and from Courtney Knauth for editorial assistance and Wynnice Pointer-Napper for design assistance.

End Notes

[1] The Office of Management and Budget issued a Statistical Policy Directive in 2008 (*Federal Register*, Vol. 73, No. 46 / Friday, March 7, 2008 / Notices), that provides guidance to Federal statistical agencies on the release and dissemination of statistical products. It stresses the need for adherence to data quality standards through equitable, policy-neutral, and timely release of information to the general public, and calls for "transparent descriptions of the sources and methodologies used to produce the data."

A Revised and Expanded Food Dollar Series   65

[2] Purchases of food by domestic institutions for people in these institutions and food purchases by domestic employers for their employees are included. Commodities such as table salt and bottled water that have no farm ingredients are excluded.

[3] Including household purchases of imported foods is necessary for the IO model to trace through the total farm sales linked to food dollar expenditures. These imported food purchases are deducted in a later step.

[4] A formal derivation of the farm- to-farm indirect matrix is provided in the appendix (see equations A29 and A30).

[5] In this context, "import-exclusive" indicates the deduction of imported food dollar purchases. The numerator in equation 3 deducts sales of only raw (farm fresh) food commodities such as imported fresh produce, and the denominator in the equation deducts all imported food dollar sales.

[6] Annual updates by BLS of the BEA 2002 benchmark IO account (aggregated to about 200 commodity/industry groups) are obtained from annual GDP data and gross industry output data, converted to chain-weighted year 2000 dollars based on consumer price index (CPI) and producer price index (PPI) statistics. An effi cient information processing algorithm know as "RAS" is used to update the benchmark technical coeffi cients for consistency with the new (survey-based) GDP and gross industry output data (www.bls.gov/emp/ ep_projections_methods.htm).

[7] "Volume" in this context is not the same as units or quantity, since different factors such as quality, production recipe, and production technologies can change over time for any given commodity.

[8] The row vector "v_m" measures the average import-inclusive value added per dollar of industry output from the employment of industry assets (e.g., hired labor, machinery, physical structures, natural resources). The value added equals 1 minus the sum of all purchased-input coeffi cients, which equals $i' L^{-1}$.

[9] Represents pre-income tax payments to capital owners, including property, plant, and equipment.

[10] Includes excise/sales/property/severance taxes, customs duties, and other fees and assessments, and deducts subsidies.

[11] Import commodity proceeds cannot be traced back to foreign labor and capital markets by industry group, so these proceeds are valued at their import prices and treated as returns to nondomestic industry assets.

[12] For example, the value of imported fertilizer is treated as import value added of the farm and agribusiness industry. The foreign natural gas industry contribution to the cost of imported nitrogen fertilizers is not reallocated to the energy industry.

[13] Math notation is as follows: matrices are denoted with capitalized letters, vectors with lowercase letters, sets with capitalized and italicized letters, and set elements (and scalars) with lower case italicized letters. Dimensions are identifi ed by subscripts in row/column order. A prime (') is the transpose operator, a '-' above a vector denotes its conversion to a diagonal matrix, and $\{\}^{-1}$ indicates a matrix inversion.

[14] The BLS accounts treat commodity imports as components of fi nal demand, entering the accounts as negative values, whereas the approach used here is to treat imports as an industry outlay (analogous to a draw-down of international inventories), and so positive values are entered.

[15] Utility outlays of the trade industries are based on the Census business expense data by 4-digit NAICS. These data are normalized to each year's BLS estimates of wholesale and retail industry utility outlays.

[16] Interindustry transactions are recorded 'free on board' (fob), and each transaction has accompanying wholesale and retail margin costs. The BLS accounts update total trade margin costs by industry, but do not itemize the margins of each transaction, so the relevant BEA benchmark margin rates are assumed and scaled up or down for each industry to replicate the updated total margin costs.

[17] Splitting the primary factor payment matrix into its positive and negative elements allows proportional scaling of both matrices to adjust positive and negative values in the same direction.

[18] For example, part of proceeds from broiler sales cover the purchase of chicks from a hatchery and animal feed from a feed manufacturer who, in-turn, purchased grain from a grain farm.

In: Consumer Food Costs
Editor: Stefanee L. Martin

ISBN: 978-1-61470-695-3
© 2011 Nova Science Publishers, Inc.

Chapter 2

HOW RETAIL BEEF AND BREAD PRICES RESPOND TO CHANGES IN INGREDIENT AND INPUT COSTS[*]

Edward Roeger and Ephraim Leibtag

ABSTRACT

The extent to which cost changes pass through a vertically organized production process depends on the value added by each producer in the chain as well as a number of other organizational and marketing factors at each stage of production. Using 36 years of monthly Bureau of Labor Statistics price indices data (1972-2008), we model pass-through behavior for beef and bread, two retail food items with different levels of processing. Both the farmto-wholesale and wholesale-to-retail price responses are modeled to allow for the presence of structural breaks in the underlying long-term relationships between price series. Broad differences in price behavior are found not only between food categories (retail beef prices respond more to farm-price changes than do retail bread prices) but also across stages in the supply chain. While farm-to-wholesale relationships generally appear to be symmetric, retail prices have a more complicated response behavior. For both bread and beef, the pass-through from wholesale to retail is weaker than that from farm to wholesale.

---

[*] This is an edited, reformatted and augmented version of the United States Department of Agriculture publication, Economic Research Report Number 112, dated February 2011.

Keywords: pass through, wholesale, retail, farm prices, beef, bread, supply chain, price transmission, price response.

SUMMARY

What Is the Issue?

Periodic spikes in the prices of major field crops and related commodities such as those from 1971 to 1974, 1994 to 1996, and 2006 to 2008 have stimulated questions about how these shocks affect wholesale and retail food prices. To what extent do wholesale food prices respond to changes in the underlying costs of inputs? How much of a change in input costs is passed through to retail prices and how long does it take for such cost changes to pass through?

Retail and wholesale prices will generally follow upstream commodity prices directionally, but there are often factors that limit this responsiveness. The extent to which price changes are passed through depends on the value added by each producer in the production process and a number of other organizational and marketing factors at each stage of production, leading to input price changes that are only partially refl ected in later stages of the supply chain, and, at times, a lack of measurable response in the downstream product's price.

In this study, we develop price pass-through models for farm-to-wholesale and wholesale-to-retail price changes using 36 years of monthly Bureau of Labor Statistics price indices data (1972-2008). We focus on the wheat to retail bread and the cattle to retail beef chains because they represent examples of supply chains with significantly different degrees of processing between stages.

What Were the Major Findings?

Pass-through rates and timing can vary dynamically between prices at different stages in the supply chain, across food categories, and for a given relationship over time.

How Retail Beef and Bread Prices Respond to Changes ... 69

- A more processed item (bread/wheat flour) showed less response to upstream price changes than did a less processed item (retail/wholesale beef).
  - o Retail beef prices typically incorporated between 19 to 29 percent of a change in the wholesale beef price after 6 months, while wholesale beef prices incorporated 52 to 54 percent of the change in cattle prices.
  - o Retail bread prices typically incorporated 16 to 21 percent of wholesale wheat flour price changes, while wholesale prices of wheat flour incorporated 29 to 31 percent of changes in wheat commodity prices.
- Wholesale prices for beef and wheat flour both responded in a generally symmetric manner to changes in farm prices regardless of the size and direction of change, while retail prices for both beef and bread adjusted asymmetrically (especially in more recent years), with the adjustment dependent upon the characteristics of the wholesale price change.
- For both beef and bread, most of the change at the farm level was passed on to the wholesale stage within the first month, with some additional adjustments to the long-term equilibrium price after that.
- Retail prices had a more complicated response to wholesale price changes, and for both bread and beef, the pass-through from wholesale to retail was weaker than the pass-through from farm to wholesale. Retail price responses were strongest when wholesale prices were relatively high. When prices were more stable or in times of price declines, signifi- cant pass-through often did not appear for several months.

How Was the Study Conducted?

We analyze the farm-to-wholesale and wholesale-to-retail price relation- ships using a two-stage error correction model that allows for the possibility of asymmetric price response. We also test for structural breaks in the long-term (cointegrating) relationships. Variations in the response of the downstream prices that are dependent on the magnitude and/or sign of changes in the upstream prices are modeled by considering a threshold-type response based on the downstream price's position relative to the expected long-term relationship.

This research extends the work of recent empirical studies that have investigated the complexity of commodity pass-through relationships using newly developed statistical tools. We characterize price-response behavior in a manner that is not overly infl uenced by any short-term market conditions that can dominate samples of fewer years by including a long time period and considering different possible types of asymmetric price adjustment. Our models also allow more freedom for the relationships between points in the supply chain to vary for a given food group and include energy and labor variables as short-term inputs.

INTRODUCTION

What is the effect of changes in commodity prices on manufacturer and retail food prices? As commodity prices surged in late 2007 and most of 2008, much focus turned to estimating the impact of these increases on retail food prices. The drop in commodity prices from late 2008 to late 2009 led to the same concerns in the opposite direction. Commodity market swings can have significant, and often complicated, impacts on retail food prices. In order to investigate these effects, we estimate how much of a change in commodity costs is passed through to retail prices, how the rate of pass-through varies by food type, and, just as important, the time lag between commodity price changes and retail price changes. From this, we gain a more detailed under-standing of the dynamic relationships between farm, wholesale, and retail prices over time and the tools to develop better expectations for the effects of farm-level price shocks on consumers. These tools will be used to refine the Economic Research Service's Consumer Price Index for Food forecasts[1] by incorporating additional farm and wholesale price changes into the forecasts.

The 2006 to 2009 upturn in agricultural price volatility was in sharp contrast to price behavior in preceding years. In measuring farm-to-retail price response behavior, such observable shifts in price volatility necessitate that flexible models be developed and utilized. To emphasize this flexibility, we focus on pricing relationships one stage at a time—the effect of farm price changes on wholesale prices, and then wholesale to retail prices—and include modeling variations that allow downstream prices to respond in a nonlinear manner. We develop a multistage model for pass-through behavior using 36 years of monthly data (1972-2008) and focus on two retail food items that have different levels of processing across their supply chains, beef and (white) bread.

Our research extends the work of some recent price transmission studies, while also focusing more closely on areas that appear absent from the current literature. We do this by looking at a time period that is longer than typical studies and we consider more flexibility in price response through the use of a model that allows for output prices to respond in a non-uniform manner to different input-price changes. In addition, we allow the relationships between points in the supply chain to vary within a food group and include energy and labor variables as short-term inputs, along with the standard food inputs. These extensions to the existing literature are included to better describe the differing nature of price response through the supply chain and across food categories.

THE PRICE SERIES DATA

Our analysis uses farm-level data on wheat and cattle prices, wholesale-level wheat flour and beef prices, and retail-level (white) bread and beef prices from 1972 through 2008.[2] In order to track general trends in these respective industries and avoid problems with following production and marketing chains for very specific retail products, we used Bureau of Labor Statistics (BLS) price indices at the aggregate product level.[3] Table 1 lists the CPI and Producer Price Index (PPI) food and commodity data series used for each product and price stage. The time-series data is monthly, not seasonally adjusted, and all price series were converted into their natural logarithms before analysis.[4] The different price series for bread and beef over this time period broadly illustrate the degree of consistency of response across price stages and changes in co-movement over time (figs. 1 and 2, respectively).

In addition to the principal food and agricultural commodity prices in each model, we also included energy and labor prices, where available. For all models we used the wholesale diesel PPI as a proxy for transport costs, while the variables for labor or additional energy inputs are more specific to the individual products and points in the supply chain. For the wholesale-to-retail pass-through relationship for both beef and bread, we controlled for variation in labor costs by using the monthly average hourly grocery store wage, while for the farm-to-wholesale relationship, we included an additional variable for the aggregate hourly slaughtering wage for beef and a variable for the electric power PPI for wheat flour.[5]

We focused specifically on white bread and beef supply chains in order to highlight differences in the degree of processing from farm to retail across these two food products. These differences can affect price-transmission relationships because as an input price represents a smaller share of the output price, it is expected that input price changes will have a smaller and/or more delayed effect on the output price. This expectation arises because in the (theoretical) example of complete price transmission, the downstream price response would be equal to the proportion of the total cost represented by the upstream price. This difference in the level of processing (or value added) to the original agricultural input can be seen in the farm share of the retail price, which in 2008 was 48 percent for beef and 10 percent for bread.

Beyond the general trends shown in figures 1 and 2, we can estimate the relative degree of price change passed through to the next stage in the supply chain by comparing price series volatility at each level in the chain. To summarize volatility, we looked at key values in the monthly price change distribution of each of the series. Table 2 presents the minimum, maximum, standard deviation, and 10th and 90th percentile values of monthly changes in each series, for the entire time period, and some selected subintervals. Within each food category (over the entire period), volatility (as estimated by the bounds of the 10th and 90th percentiles) decreased when moving from farm-to-retail prices, showing that downstream prices are more stable and price response is decreasing through the distribution chain. Between food categories, bread and wheat flour prices are less volatile than beef prices

at the retail and wholesale stages. The differences among food categories' price volatility probably results from the higher degree of processing (which implies less use and reliance on the agricultural input commodity) for wheat flour and bread that has led to fewer price swings and less pass-through of the wheat price volatility.

Table 1. Time series price variables

Supply chain level	Bread	Beef
Retail	White bread CPI	Beef and veal CPI
Wholesale	Wheat fl our PPI	Beef and veal, fresh or frozen PPI
Farm	Wheat PPI	Cattle PPI

CPI = Consumer Price Index; PPI = Producer Price Index Source: USDA, Economic Research Service.

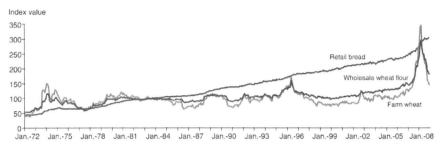

Source: USDA, Economic Research Service calculations based on U.S. Department of Labor, Bureau of Labor Statistics data.

Figure 1. Bread price indices at different production stages.

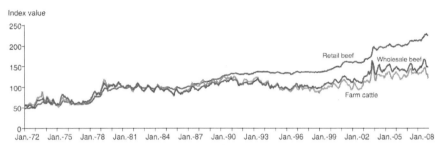

Source: USDA, Economic Research Service calculations based on U.S. Department of Labor, Bureau of Labor Statistics data.

Figure 2. Beef price indices at different production stages.

Table 2 also shows that within shorter time periods, the amount of price variation can vary dramatically from the overall time-period average. This range of price variations has implications for studies that try to quantify pass-through rates from a limited time horizon, since these studies may find results that are due to a particular pricing environment and may not be representative of other situations or time periods. For example, if considering the co-movement between wholesale wheat flour and retail bread prices in the plot of the price series from the late 1990s through the early 2000s, it might seem that there is very little price response from wholesale to retail (see fig. 1). With this focus, the more-than-15-percent increase in the retail bread CPI over the short period from mid-2007 to mid-2008 would look unprecedented.[6] The years included in our analysis were chosen with the goal of including as much dynamic movement in the price series as possible under the constraint of having consistently available data for all of the variables in the model. There is, however, a tradeoff with using a sample covering a large number of years in that the possibility of structural change in relationships over time becomes more prevalent. As will be described later, steps are taken in our analysis to address this issue.

Table 2. Price series volatility measures for different products in different time periods

Variable	Time period	10th and 90th percentiles		Standard deviation *Percent*	Largest decrease	Largest increase
Δ White bread CPI	1972-2008	-0.66	1.44	0.97	-2.11	8.18
	1990-2008	-0.92	1.49	0.96	-2.11	3.52
	2000-2006	-1.03	1.40	0.96	-2.11	2.64
Δ Wheat flour PPI	1972-2008	-3.70	3.88	3.67	-19.12	22.87
	1990-2008	-3.78	4.36	3.53	-11.39	15.67
	2000-2006	-1.99	3.03	2.04	-4.76	6.01
Δ Wheat PPI	1972-2008	-7.33	7.27	6.96	-25.51	63.85
	1990-2008	-7.77	7.78	6.53	-25.51	22.02
	2000-2006	-6.12	8.93	5.87	-12.61	17.43
Δ Beef CPI	1972-2008	-1.10	1.82	1.59	-5.62	7.36
	1990-2008	-0.66	1.27	0.94	-2.39	7.15
	2000-2006	-0.66	1.51	1.24	-2.39	7.15
Δ Beef PPI	1972-2008	-4.23	4.47	3.90	-13.70	18.10
	1990-2008	-3.64	3.51	3.16	-11.24	15.14
	2000-2006	-3.88	4.10	3.73	-11.24	15.14
Δ Cattle PPI	1972-2008	-4.79	5.34	4.53	-19.25	19.64
	1990-2008	-4.06	4.44	3.97	-19.25	18.55
	2000-2006	-3.98	5.53	4.82	-19.25	18.55

Note: The 10th and 90th percentiles represent the range of numbers that are the bounds that 80 percent of the monthly changes fall between. CPI = Consumer Price Index; PPI = Producer Price Index

Source: USDA, Economic Research Service calculations based on U.S. Department of Labor, Bureau of Labor Statistics data.

Long-Term Price Relationships
and Structural Breaks

A basic pass-through relationship between price series in a supply chain relates an output price to an input price by viewing the downstream product as essentially a value-added version of the upstream product. However, the amount of "value" that is added and the inclusion of other inputs can have substantial effects on the price response of the retail food product to changes in its principal agricultural input's price. Over a long enough time horizon, changes to this difference between downstream and upstream price series can significantly affect the price response.

The farm share of the retail food dollar for beef as well as cereals and bakery products has been declining over time due to the increased demand for and supply of value-added convenience items in both of these categories. This trend is not limited to beef and bread. Additional processing and food preparation beyond the farmgate has increased the number of ready-to-eat products available to consumers and decreased the farm share for all food products from 32 percent in 1970 to 19 percent in 2006. For beef and cereals and bakery categories, specifically, the farm share dropped from 64 to 46 percent and 16 to 10 percent, respectively, during that time period.[7] As these numbers show, retail bread prices have typically had a lower share of input commodity prices than beef, but there has been a significant decline in the farm share of retail beef prices over the last 38 years. Such changes have certainly affected price pass-through rates between points in the supply chain.

Given that changes in pass-through may occur over time, we begin with a model that allows for the relationship between input and output prices to vary in different sub-intervals of time. We therefore express the long-term relationship between two prices as:

$$P_{O,t} = \beta_0 + \beta_1 P_{I,t} + \beta_2 \varphi_1 + \beta_3 \varphi_2 + \beta_4 \varphi_3 + u_t. \tag{1}$$

where $P_{O,t}$ and $P_{I,t}$ represent the output and input price series at time t, respectively, $\beta_0 - \beta_4$ are parameters to be estimated, φ terms represent time-period specific dummy variables, and u is an error term. The φ terms represent structural-break variables and provide time-sensitive measures of differences in the long-term relationship between P_O and P_I, which are points in time at which the relationship between P_O and P_I is diverging (assuming $\beta_2, \beta_3, \beta_4 > 0$).

In order to let the data drive the specification of these time-period specific dummy variables, we follow an approach similar to Boetel and Liu (2008) in their investigation of the longrun price linkage between farm, wholesale, and retail beef and pork prices. Rather than imposing assumptions on the data regarding when structural breaks occur, we explore patterns within the data in order to identify potential structural breaks endogenously. We find three structural break dates during the 36 years in each of the wholesale-to-retail price relationships and two in each of the farm- to-wholesale relationships (table 3).[8] Plots of the price indices with markers for the estimated break dates are given in figures 3 and 4, and the estimates of the long-term equations with structural breaks are presented in table 4.

Table 3. Estimated structural breaks in long-term relationships

	Supply chain relationship	Estimated break dates
Beef		
	Wholesale to retail	Oct. 1980, June 1991, June 2001
	Farm to wholesale	April 1995, April 2000
Bread		
	Wholesale to retail	March 1980, July 1989, May 1997
	Farm to wholesale	May 1983, March 1998

Source: USDA, Economic Research Service calculations based on U.S. Department of Labor, Bureau of Labor Statistics data.

As expected, within a food category, the break dates for different price stages occur at similar times. The estimated coefficients corresponding to the go terms provide some information as to how the long-term relationships between these upstream and downstream prices have changed since the 1970s. That is, the magnitudes consistently grow larger with the later structural breaks (i.e., $/\beta_2 < /\beta_3 < /\beta_4$), confirming that output prices have been diverging from input prices over time. This trend is much more pronounced for retail prices and is especially strong for retail bread prices.

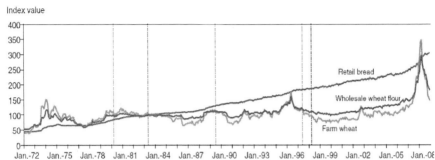

Note: Red dotted lines represent estimated structural breaks in the wholesale-to-retail relationship. Black dotted lines represent estimated structural breaks in the farm-to-wholesale relationship.

Source: USDA, Economic Research Service calculations based on U.S. Department of Labor, Bureau of Labor Statistics data.

Figure 3. Bread price indices at different production stages.

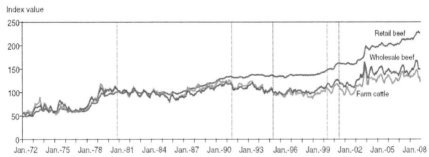

Note: Red dotted lines represent estimated structural breaks in the wholesale-to-retail relationship. Black dotted lines represent estimated structural breaks in the farm-to-wholesale relationship.

Source: USDA, Economic Research Service calculations based on U.S. Department of Labor, Bureau of Labor Statistics data.

Figure 4. Beef price indices at different production stages.

While shifts in the relationship among farm, wholesale, and retail prices over the last 36 years may not seem surprising, it is helpful to consider the background of these shifts in more depth. As previously mentioned, the farm shares for both beef and bread fell considerably over the period, implying that other factors have gained in significance over time. Hahn (2004) explores several reasons behind increasing (nominal) farm-to-wholesale and wholesale-to-retail price spreads. He finds that increasing productivity in the meatpacking and livestock industries have lowered real farm and wholesale

78 Edward Roeger and Ephraim Leibtag

prices (and the infl ation-adjusted price spread) from 1970 to 2003, while an expanding service component in grocery stores has increased gross real margins between wholesale and retail meat values. Assuming similar trends in the bread supply chain (agricultural and processing productivity increases while overall grocery-store productivity falls) helps explain why we find more breaks in the wholesale-to-retail relationships than in the farm-to-wholesale relationships, as well as larger coefficients (implying faster growing margins) on the wholesale-to-retail break variables.

Table 4. Long-term relationship estimates with structural breaks

	Supply chain relationship	Estimated equation
Bread		
	Wholesale to retail Farm to wholesale	$P_{O,t} = 1.983 + 0.492\,P_{I,t} + 0.404_{\varphi1} + 0.733_{\varphi 2} + 1.024_{\varphi3}$, $R^2{=}0.96$ $P_{O,t} = 1.385 + 0.676\,P_{I,t} + 0.176_{\varphi 1} + 0.305_{\varphi 2}$, $R^2{=}0.96$
Beef		
	Wholesale to retail Farm to wholesale	$P_{O,t} = 0.413 + 0.882\,P_{I,t} + 0.198_{\varphi1} + 0.399_{\varphi 2} + 0.496_{\varphi3}$, $R^2{=}0.98$ $P_{O,t} = 0.627 + 0.859\,P_{I,t} + 0.074_{\varphi 1} + 0.158_{\varphi 2}$, $R^2{=}0.98$

Notes:

All estimated coefficients were statistically significant at an error rate of < 0.01 percent.

The variable $_{\varphi 1} = 1$ for Break Date 1 $\leq t <$ Break Date 2, otherwise $_{\varphi 1} = 0$.

The variable $_{\varphi 2} = 1$ for Break Date 2 $\leq t <$ Break Date 3, otherwise $_{\varphi 2} = 0$.

The variable $_{\varphi 3} = 1$ for t > Break Date 3, otherwise $_{\varphi 3} = 0$

Refer to table 3 for the estimated break dates for each model.

Source: USDA, Economic Research Service calculations based on U.S. Department of Labor, Bureau of Labor Statistics data.

Significant specific supply-and-demand changes also have occurred across these industries through the 1972-to-2008 sample time period. In the bread supply chain, a long-term trend in increasing acreages of wheat planted in the United States ended in 1981 with acreages since then dropping off considerably (Ali, 2002). Trends in consumption also changed. In 1997, per capita wheat-fl our consumption began to decline after steadily increasing since the 1970s. For beef, production shifts occurred over the sample period leading to increased grower-operation packer sizes and increased industry concentration. For example, the share of purchases made by the four largest beef processors doubled between 1980 and 1990, and significant increases in operation sizes also occurred between 1992 and 1997. The production locus[9] for cattle-raising operations increased from 23,891 head to 38,000 head for fed cattle from 1992 to 1997 (MacDonald and McBride, 2009). This increase in production locus was more than twice the increase in size from 1987 to 1992. Toward the end of the 1972-to-2008 sample period, the Congressional Livestock Mandatory Reporting Act of 1999 was implemented by USDA in 2001. From 1997 to 2002, the trend of consolidation and rapid growth in beef-cattle operation size leveled off.

A MORE DETAILED MODEL FOR PRICE TRANSMISSION

Equation 1 accounts for the long-term relationship between a downstream and upstream price series, but in describing pass-through, we are interested in the short-term dynamics as well, in order to more completely explain how a change in one price will be refl ected in the change of another price series. Equation 1 should not be disregarded—to the contrary, the equation can be useful as an estimation of how the actual P_O compares to its expected value as predicted by the long-term relationship. To better capture the full pass-through relationship, we use an error correction model (ECM) that includes measures of short-term changes as well as adjustments to the expected longterm relationship, and is expressed as:

$$\Delta P_{O,t} = \alpha_0 + \Sigma^Q_{i=1} (\alpha_{1,i} \Delta P_{O,t-i}) + \Sigma^R_{i=1} (\alpha_{2,i} \Delta P_{I,t-i}) + \Sigma^S_{i=1} (\alpha_{3,i} \Delta x_{1,t-i})$$
$$+ \Sigma^T_{i=1} (\alpha_{4,i} \Delta x_{2,t-i}) + \gamma u_{t-1} + \upsilon_t,$$
$$(2)^{10}$$

where x_1 and x_2 are variables that are assumed to have an effect on P_O in the short-term without necessarily having a stable long-term relationship with it, and υ is the residual from the ECM. In our analysis, the ECMs are constructed following Engle and Granger (1987). The constant term, α_0, and dummy variables corresponding to the identified structural break dates may be included conditional on the output price series appearing to have a clear trend over time; energy and labor inputs are modeled as short- term variables, in that they are present in the error correction model but not in the long- term equation. This is an ECM because of the γu_{t-1} term that represents changes in P_O due to the previous period's value of u (which is the part of P_O that is unexplained by the other terms in equation 1). This particular model is a symmetric ECM because the response of $\Delta P_{O,t}$ is the same regardless of the magnitude and sign of the $\Delta P_{I,t-i}$ and u_{t-1} terms. That is, input price increases are passed on to output prices as completely and quickly as input price decreases.

In recent years, many empirical studies have investigated the complexity of commodity pass-through relationships using this relatively new methodology that incorporates both short- and long-term relationships through ECMs. Goodwin and Harper (2000), for example, combine an ECM with the possibility of a nonlinear threshold setting in studying weekly pork prices from 1987 to 1999.[11] They find evidence that retail prices respond to upstream price changes differently depending on behavior characterized by regimes that are defined by different threshold values. Boetel and Liu (2008) also consider an ECM with a focus on livestock pricing. Looking at a longer time period (1970 to 2008), they investigate price response in light of structural breaks in the long-term relationships between prices across the supply chain. Both of these studies find it beneficial to model the pass-through relationship as a combination of (1) a short-term response to input price changes and adjustments to an expected long-term equilibrium and (2) asymmetric price responses that allow output prices to respond differently depending on the direction of input price changes.

ASYMMETRIC PASS-THROUGH BEHAVIOR

Inspection of the graphs of the different bread and beef price levels (figs. 1 and 2) shows that while the downstream price (in most cases) does seem to have a tendency to follow changes in the upstream price, this behavior is not always consistent across all changes. Wholesale prices generally follow farm price changes fairly close, but for retail bread prices in particular, large responses to upstream price changes seem to occur only infrequently (especially in the last 30 years). With retail bread prices, only very substantial changes in wheat flour prices seem to elicit a response.

An asymmetric price response is defined as a relationship in which the output price does not necessarily respond proportionally to all input price changes, but instead varies depending on either the magnitude or the sign of the change in input prices. Why might price transmission be dependent upon the magnitude and sign of the input price change? Awokuse and Wang (2009) cite some possible theories that may result in asymmetric price transmission,

How Retail Beef and Bread Prices Respond to Changes ...

including noncompetitive market structures, price rigidity due to transaction costs, and commodity storage characteristics. A review of a number of works focused on the underlying theories behind asymmetric price transmission by Meyer and von Cramon-Taubadel (2004) finds that, for the direction of asymmetry, there are arguments for either increases or decreases to cause greater downstream responses depending on the specific circumstances of the industry in question. The authors also point out (page 582) that, "Existing tests describe the nature of price adjustment but most are not discerning in the sense that they make it possible to differentiate between competing underlying causes on the basis of empirical results." Kinnucan and Forker (1987) also suggest that, even if a retail price responds symmetrically in the long run, delays may arise that increase the response time. They cite issues such as normal marketing inertia, repricing costs, and differences in information collection and transmission as all working to slow down or mitigate price transmission. Taken together, such factors can lead to incomplete pass-through across the supply chain and, at times, a lack of measurable response in the downstream product's price.

In this study, we allow output prices to respond asymmetrically to both adjustments in the short-term price response or corrections to the long-term relationship by using a threshold model. This allows for price responses to vary depending on certain threshold values that act as bounds to different pass-through behaviors. Using this model leads to the following transformation of equation 2:

$$\Delta P_{O,t} = \{f^{(1)}(\Delta P_{O,t-i}, \Delta P_{I,t-i}, \Delta x_{1,t-i}, \Delta x_{2,t-i}, u_{t-1})\} \qquad \text{if} \quad u_{t-1} \leq c_1,$$
$$\{f^{(2)}(\Delta P_{O,t-i}, \Delta P_{I,t-i}, \Delta x_{1,t-i}, \Delta x_{2,t-i}, u_{t-1})\} \qquad c_1 < u_{t-1} \leq c_2,$$
$$\{f^{(3)}(\Delta P_{O,t-i}, \Delta P_{I,t-i}, \Delta x_{1,t-i}, \Delta x_{2,t-i}, u_{t-1})\} \qquad u_{t-1} > c_2 \tag{3}$$

in which $f^{(1)}$, $f^{(2)}$, and $f^{(3)}$ all have the same general form, essentially equation 2 (with the possibility of different lag lengths).[12] The terms c_1 and c_2 refer to the lower and upper threshold bounds, respectively[13] and are in terms of values of the variable u_{t-1} (the difference between the actual P_O and its expected value from equation 1). This variable is used because it represents a comparison of the downstream price relative to its long-term expectation, and the threshold bounds are constrained such that $c_1 < 0$ and $c_2 > 0$. The first grouping of observations by the thresholds, or regime, are points in time in which the output price is relatively low compared with what is expected from the estimated long-term relationship ($u_{t-1} \leq c_1$), the second regime is for observations in which the output price is relatively consistent with the long-term expectation ($c_1 < u_{t-1} \leq c_2$), and the third regime is when the output price is relatively high ($ut\text{-}1 > c_2$). By breaking up the estimation of equation 2 for each of these different regimes, pass-through rates are allowed to differ depending on the deviations in the current relationship of P_O and P_I from the expected long-term relationship between these input and output prices. The sign *and* magnitude of input price changes may lead to different output price responses in this threshold model because, for example, if P_O is relatively close to its expected value, then (holding P_O constant) a large increase (decrease) in P_I will result in a large negative (positive) u value and categorization to the first (third) regime, while a small change in P_I will result in a u value that is small in magnitude and categorization in regime 2.

Figures 5-8 show the threshold values, the number of observations in each regime, and the patterns of u values (deviations from the expected long-term relationships) over time. For both the beef and bread categories, the values of the thresholds themselves can also be descriptive. The bounds for the beef threshold models are more symmetric around zero, which implies that for the threshold wholesale-to-retail beef model, the regimes are more clearly defined as large positive and negative deviations from the long-term relationship (regimes 1 and 3, respectively) or generally small deviations (regime 2).

The bounds for both the wheat flour and bread threshold models, however, are not as symmetric around zero and the relatively smaller upper bound implies that, in these cases, the middle regime will be more balanced toward observations in which the expected downstream price is relatively low.

As an example of the mechanics of a threshold model, consider how a threshold ECM fits the price data for retail beef prices over a 12-month period and how the different pass-through estimates from each regime can provide a better fit at different points in time (fig. 9). When retail prices are more responsive (which our threshold ECM finds to be generally when retail prices are relatively low compared with the long-term relationship), the first regime estimates, which have higher pass-through rates and stronger error correction, mimic the actual response better. Conversely, there are also settings in which actual retail beef prices are less responsive and the pass-through estimates of the first regime would overpredict volatility in retail prices. In most of these cases, the threshold ECM applies the lower pass-through estimates of the second or third regimes (in which retail prices are about in line with long-term expectations or they are high compared with wholesale prices), and the predicted responses from the threshold ECM more closely follow the actual retail beef responses.

How Retail Beef and Bread Prices Respond to Changes ... 83

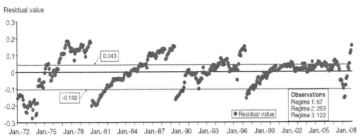

Source: USDA, Economic Research Service calculations based on U.S. Department of Labor, Bureau of Labor Statistics data.

Figure 5. Bread, retail-wholesale thresholds and long-term relationship residual values Residual value.

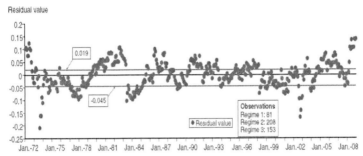

Source: USDA, Economic Research Service calculations based on U.S. Department of Labor, Bureau of Labor Statistics data.

Figure 6. Bread, wholesale-farm thresholds and long-term relationship residual values Residual value.

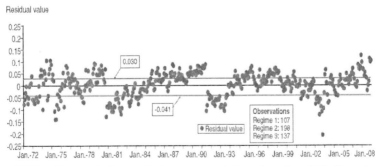

Source: USDA, Economic Research Service calculations based on U.S. Department of Labor, Bureau of Labor Statistics data.

Figure 7. Beef, retail-wholesale thresholds and long-term relationship residual values Residual value.

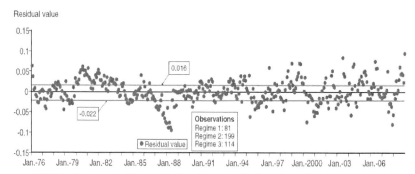

Source: USDA, Economic Research Service calculations based on U.S. Department of Labor, Bureau of Labor Statistics data.

Figure 8. Beef, wholesale-farm thresholds and long-term relationship residual values.

As noted earlier, a threshold model allows for price responses to differ based on the magnitude and/or sign of the input price change. Thus, this model will describe pass-through relationships more accurately and fit the data more closely when the downstream price does have a tendency to respond to input price changes in an inconsistent manner. For some food categories and stages in the supply chain, under certain conditions, marketing inertia causes downstream prices to be infl exible or unresponsive. Other categories and stages are less likely to experience such marketing inertia.

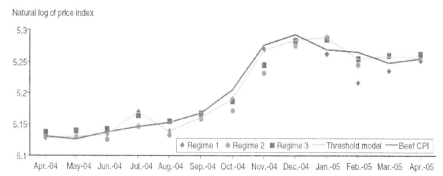

CPI= Consumer Price Index.

Source: USDA, Economic Research Service calculations based on U.S. Department of Labor, Bureau of Labor Statistics data.

Figure 9. Comparison of actual observations and threshold model by regimes.

In our analysis, two different measures point toward a threshold setting as a good fit for the wholesale-to-retail price relationships but not for the farm-to-wholesale price relationships. The first measure is a statistical test that seeks to confirm the significance of threshold effects with the chosen threshold values.[14] The second measure compares predictions made for the change in a downstream price using a basic ECM (as in equation 2) and a threshold ECM and then builds on this prediction for a total of six consecutive monthly predictions. Table 5 highlights the findings of this application for a sequential series of 6-months-ahead predictions with starting points in each month from 2002 through 2008, showing the average prediction error by model (averaged across each 6-months-prediction horizon and then across the entire series of these predictions). The ranking of the results for each model indicates that the threshold models do not perform better than the symmetric ECMs in the farm-to-wholesale stages but are preferred in the wholesale-to-retail models. Thus, our findings indicate that for both wholesale beef and wholesale wheat flour, pass-through of farm level price changes appears to occur in a fairly uniform manner regardless of the size and direction of the change. For retail beef and bread prices, the response to an input price change may differ significantly depending on the magnitude and sign of the change.

Table 5. Six-month prediction comparison for 2002-08[1]

	Supply chain relationship	Average forecast error[2]
Beef		
Wholesale to retail	Threshold ECM Symmetric ECM	1.6689 2.0470
Farm to wholesale	Threshold ECM Symmetric ECM	4.3210 4.2752
Bread		
Wholesale to retail	Threshold ECM Symmetric ECM	2.1001 2.3680
Farm to wholesale	Threshold ECM Symmetric ECM	7.2701 7.2350

[1] Across the period 2002 to 2008, the different models were used to make 6-months-ahead predictions using each month as a different starting date.

[2] This can be described as the mean forecast error for each 6-months-ahead prediction horizon, averaged across the entire series of these predictions.

ECM = Error correction model.

Source: USDA, Economic Research Service calculations based on U.S. Department of Labor, Bureau of Labor Statistics data.

PASS-THROUGH MODEL ESTIMATES

We estimate each of the pass-through ECMs—symmetric (equation 2) and threshold (equation 3)—for each supply-chain relationship for beef and bread with a flexible lag order across models and regimes (appendix tables A1-A4).[15] Although estimates are given for both threshold and symmetric models in all cases, our discussion focuses on the symmetric ECMs for farmto-wholesale movement and the threshold (asymmetric) ECMs for wholesaleto-retail movement, following our earlier discussion of model ft.

We f rst look at the beef farm-to-wholesale symmetric ECM results as an example of how to interpret the estimated coeff cients. A coeff cient of 0.34 on Δ *(ln cattle PPI)$_{t-1}$* implies a direct pass-through rate of 34 percent of a price change in the cattle price index to wholesale beef prices after 1 month. Also, the estimated coeff cient of the error correction term (ECT) of -0.14 implies that there is some adjustment based on the difference between the last month's actual wholesale beef price and its expected value (as predicted by the long-term relationship from equation 1). For the ECT estimates, the magnitude of the number corresponds to the speed of adjustment to the long-term relation-ship, while a negative (positive) sign implies convergence (divergence) to the long-term relationship. The ECT estimates are most easily interpreted in a relative rather than direct manner. For example, the estimated ECTs for whole-sale beef of -0.14 and wholesale wheat flour of -0.07 both imply that there is pressure on the respective prices to converge to the long-term relationship, but the larger magnitude of the estimate for beef suggests that the effect is stronger there (and thus adjustment to the long-term relationship is faster).

Several patterns emerge between products and between price stages from these regression results. Between food products, the strength of the pass-through rate is inversely correlated with the level of processing of the input commodity, thus beef generally has larger and quicker pass-through than bread/wheat flour. This can be seen in the direct pass-through responses (the $\Delta P_{I,t-i}$ terms, where I is the agricultural input price) which, in the f rst instance of direct response, for retail and wholesale beef models are 0.13 to 0.27 and 0.34, respectively.[16] This is in contrast to 0.05 to 0.10 and 0.12 for retail bread and wholesale wheat flour, respectively. When looking at these numbers across the supply chain instead of across food categories, the farm-to-wholesale price relationships also show more direct pass-through than that of wholesale-to-retail prices.

Looking now at farm-to-wholesale (symmetric ECM) results in more detail, we f nd that wholesale beef prices have a strong and immediate response to cattle price changes with pass-through comprised of a direct response after 1 month and strong error correction to the long-term relationship. Both of the coeff cients for these responses, 0.34 and -0.14, respectively, are the largest coeff cients estimated in any of the models in this study. These coeff cients highlight the close co-movement of the two series, even in recent years.

For the effect of wheat price changes on wheat flour, the symmetric ECM describes the response as quick yet relatively modest. One reason for this seemingly low short-term response rate of 11.5 percent after 1 month (with statistically signif cant but relatively modest error correction of -0.07, as well) is that wheat prices are generally prone to relatively large temporal swings, while wheat flour prices are generally more stable, implying less of these input changes are passed through.

Retail prices have a more complicated response behavior than wholesale prices, but for both food products the pass-through at this stage is weaker than the upstream stage. The retail bread threshold ECM estimates show pass-through to be strongest and fastest (10 percent directly after 1 month and -0.12 for the ECT) for the first regime, which characterizes times when retail prices are much below what would be predicted by the long-term relationship. When prices are approximately in line with long- term expectations or input prices are slightly increasing (regime 2), there is an estimated response of about 11 percent after a delay of 3 to 4 months and some slight response from the ECT (-0.03). When wheat flour prices are rapidly falling or when retail prices are relatively high (regime 3), wheat flour price changes still have an effect (about 15 percent pass-through delayed between 2 and 4 months), but the retail prices have no significant tracking to the long-term relationship between the series.

Beef retail prices seem to follow a similar pattern. Retail price response is strongest when wholesale prices are surging (first regime) with 38 percent direct pass-through within 2 months and -0.12 for the ECT. In times of modest changes or when retail prices are relatively high (regimes 2 and 3), there is still a fairly high level of responsiveness (about 22 and 31 percent, respectively) within a couple of months. Also, in this model, retail beef prices have significant adjustment back to the long-term relationship only after large wholesale price increases (while, in the second regime, the positive value of the ECT actually has a somewhat divergent effect between the series).

There were also some similarities in results across stages in the supply chain for the nonagricultural input prices.[17] For both food categories, the wholesale price response to diesel price changes is small in magnitude but significant and fairly quick. The other input variables (slaughtering wage and electricity) were both significant in their respective models and had effects occurring with a much greater lag as compared with diesel prices. In the wholesale-toretail models, only retail beef prices had a significant response to labor or energy price fluctuations with changes to the grocery store wage appearing to have a modest effect after 2 months.

SUMMARIZING THE EXPECTED PASS-THROUGH RATES

Interpreting the full results of a nonlinear model such as the threshold ECM can be difficult since the pass-through rate depends upon the sign and magnitude of the input price change as well as the time period in question.[18] A tool that can be helpful in exploring the results of these types of models is a nonlinear impulse response function (NLIRF).[19] The NLIRF is a simulation approach that can be used to gauge the impact of a specific change at a specific point in time. Beyond being able to focus on a particular point in time, this method combines the total expected pass-through from both the short-term response and error correction to the long-term relationship.

To summarize the pass-through results for the different models and provide some measure for response that is inclusive of the different pass-through factors, we calculated the cumulative short-term pass-through coefficients, timing for these coefficients, and estimates for pass-through based on NLIRF results for a 6-month time span (table 6). The first two rows of each section of the table summarize the information contained in the appendix tables. The other results in this table (NLIRF results, described as "6 Month Total") are presented to give a more complete sense of the pass-through that combines the different effects of the ECMs—the direct pass-through, adjustment to the long-term relationship, and any regime switching in the threshold models. We used a one-standard-deviation input price change (positive or negative) in our simulation in order to present a characteristic response.[20] The results presented here are a summarized average of the percent of the input price change that was passed through after 6 months, using each month from January 2000 through January 2008 as starting points.[21] The values for the NLIRF results lead to conclusions similar to those discussed in the previous section. On average from 2000 through 2008, retail and wholesale beef prices are more responsive to input price changes (19.2 to 28.6 percent and 52.6 percent, respectively) than are retail bread and wholesale wheat flour prices (16.3 to 21.4 percent and 30.3 percent, respectively). Comparing the same numbers across the supply chain, wholesale price responses are generally stronger than retail price responses.

As the threshold regression results for the wholesale to retail relationships showed, estimated pass-through rates are significantly different among regimes, with Regime 1 (relatively low retail prices) having the highest pass-through rates for both beef and bread. When interpreting the NLIRF results of table 6, it is important to consider that the pass-through rates from the threshold models are sensitive not just to the value of u (deviations from the expected long-term relationship), but also to how close the u value is to the threshold bounds. The relative position of u is important for these threshold models because it determines which regime a time period falls into and the amount of change necessary to switch regimes. This movement among regimes describes how the overall pass-through behavior of the output price changes with respect to the value of u because the estimated amount of pass-through and the timing differ between regimes.

From 2000 through 2008, retail bread prices typically appeared to be relatively high (in terms of the u variable) and close to the Regime 3 boundary, while retail beef prices typically were also in Regime 2 but in a less concentrated pattern. This difference likely explains why the models predict that retail bread prices, on average, will respond more to an input price decrease than to an input price increase of similar magnitude. Retail bread prices were much more clustered around the boundary for switching to Regime 3 (retail prices being relatively high) than were retail beef prices. This is evidenced by the finding that, for beef, a 6.7-percent wholesale price increase would generally move an observation from Regime 2 into Regime 1 (in which pass-through rates are estimated to be higher and retail prices are relatively low) or a 4.1-percent wholesale price decrease for moving from Regime 2 to Regime 3. The same sets of numbers for bread are 25.2 percent and 6.3 percent, respectively. This implies that a very large input price increase was generally needed to have higher pass-through rates (Regime 1) in retail bread prices while a relatively modest wheat flour price decrease would lead to the slightly higher pass-through rates estimated for Regime 3.

Table 6. Pass-through summary

| | Threshold ECM | | | Symmetric ECM |
	Regime 1	Regime 2	Regime 3	ECM
Beef				
Wholesale to retail				
Total direct response[1]	38.0	31.0	19.6	31.2
Timing[2]	1 - 2	1	1 - 2	1 - 2
6-month total[3], $\Delta P_I > 0$		28.63		
6-month total, $\Delta P_I < 0$		19.22		35.1
Farm to wholesale				
Total direct response	28.7	41.1	22.3	34.0
Timing	1	1	1	1
6-month total, $\Delta P_I > 0$		48.03		
6-month total, $\Delta P_I < 0$		47.66		52.6
Bread				
Wholesale to retail				
Total direct response	10	10.8	15.5	18.5
Timing	1	3 - 4	2 - 4	1 - 4
6-month total, $\Delta P_I > 0$		16.32		
6-month total, $\Delta P_I < 0$		21.38		18.7
Farm to wholesale				
Total direct response	11.2	26.3	19.3	11.5
Timing	5	1 - 2	1	1
6-month total, $\Delta P_I > 0$		30.22		
6-month total, $\Delta P_I < 0$		39.68		30.27

[1]Total direct response refers to the cumulative direct pass-through (percentage) without considering the effect of the long-term relationship between price series.

[2]This is the range of months (after the input price change) that the direct pass-through is present.

[3]This is the cumulative pass-through (percent) (with error correction) after 6 months for an impulse of 1 standard deviation of change, and the average of using each month in the period January 2000 to January 2008 as a different starting date.

ECM = Error correction model.

Source: USDA, Economic Research Service calculations based on U.S. Department of Labor, Bureau of Labor Statistics data.

Although the results in table 6 provide characteristic responses for a given period of time, we were also interested in considering how the threshold models for retail prices perform at specific points in time when markets are stressed and pass-through rates may be higher. Figures 10 and 11, therefore, show NLIRF examples for retail beef and bread price responses on a month-by-month basis for 1 and 3 standard deviation changes in the downstream price, when downstream prices were accelerating at an above-normal rate. For an input price increase (decrease) of one standard deviation, total pass-through is estimated to be 63.0 percent (36.1 percent) for retail beef and 37.4 percent (15.5 percent) for retail bread. This is in contrast to the lower pass-through rates in table 6 because at these times rapid input price increases trigger price-response behavior in the first regime as the slower adjusting retail prices are especially low compared with the expected long-term relationship. Thus, pass-through rates can be highly variable and dependent upon the relative relationship between price series.

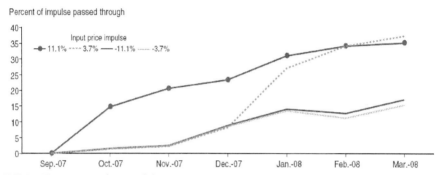

ECM = Error correction model.
Source: USDA, Economic Research Service calculations based on U.S. Department of Labor, Bureau of Labor Statistics data.

Figure 10. Pass-through in the nonlinear impulse response function bread wholesale-retail, threshold ECM.

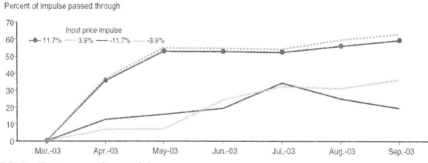

ECM = Error correction model.
Source: USDA, Economic Research Service calculations based on U.S. Department of Labor, Bureau of Labor Statistics data.

Figure 11. Pass-through in the nonlinear impulse response function beef wholesale-retail, threshold ECM.

CONCLUSION AND FUTURE EXTENSIONS

Our results indicate that wholesale prices respond to changes in farm-level prices in a generally symmetric manner, with the largest price response occurring within 1 month and some additional pass-through from adjustments to the long-term relationship after that. Single-period pass-through response estimates also were generally higher for farm to wholesale than wholesale to retail. Retail price responses to wholesale price changes are characterized by more complex behavior with threshold effects that are statistically significant and direct pass-through of changes at times occurring quickly (within 1 month) and under other conditions more slowly (taking between 2 to 4 months). Differences between food categories also exist, with more processed items (bread and wheat flour) showing less response to upstream price changes than less processed items (retail and wholesale beef).

Although the results of our study are robust to a number of different model specifications, there is an implicit assumption in our models that the direction of response between prices in the supply chain follows a single, natural path from farm to wholesale to retail and that downstream prices have negligible or inconsistent feedback on their upstream prices. This assumption stems from both the more direct and (the assumed) stronger effect of an input price on its output price (than vice versa) and the common finding that retail-price changes having little impact back through the supply chain to commodity-price changes.[22]

By modeling two food categories and two pricing relationships within each supply chain, our study provides examples of pass-through analyses, but there are a number of extensions to our work that would enhance understanding of pricing behavior in the food marketing system. One would be to conduct an analysis using price measures that are available with greater frequency. This may provide a useful comparison to our results, but such data are unlikely to include more than a few years of observations. Another path of inquiry would be to consider more points in the supply chain for some food categories in order to trace price change linkages even further back in the production chain. An example of this would be to trace back the effect of corn and soybean price changes on cattle prices, and thus gain further insight to how basic commodity prices affect retail markets.

Aside from these avenues, the most basic continuation of this work would be the application of these types of models to other food categories. This endeavor would be quite useful because, while similarities among groups are likely to exist, pass-through behavior itself is unique to each input and output relationship. Once these additional food categories are modeled, an update to ERS's forecasting of the Food Consumer Price Index and its subcomponents could be implemented and should improve our ability to predict changing trends in retail food-price inflation.

REFERENCES

Abdulai, Awudu. 2002. "Using Threshold Cointegration to Estimate Asymmetric Price Transmission in the Swiss Pork Market," *Applied Economics*, Vol. 34, pp. 679-687.

Ali, Mir. 2002. "Characteristics and Production Costs of U.S. Wheat Farms," SB-974-5, USDA, Economic Research Service. Available at: http://www.ers.usda.gov/publications/sb874-5/.

Awokuse, Titus, and Xiaohong Wang. 2009. "Threshold Effects and Asymmetric Price Adjustments in U.S. Dairy Markets," *Canadian Journal of Agricultural Economics*, Vol. 57, pp. 269-286.

Balke, Nathan, and Thomas Fomby. 1997. "Threshold Cointegration," *International Economic Review*, Vol. 38, No. 3, pp. 627-645.

Boetel, Brenda, and Donald Liu. 2008. *Incorporating Structural Changes in Agricultural and Food Price Analysis: An Application to the U.S. Beef and Pork Sectors*, Working Paper 2008-02, The Food Industry Center, University of Minnesota.

Elliot, G., T Rothenberg, and J. Stock. 1996. "Efficient Tests for an Autoregressive Unit Root," *Econometrica*, Vol. 64, pp. 813-836.

Engle, Robert, and Clive Granger. 1987. "Co-Integration and Error Correction: Representation, Estimation, and Testing," *Econometrica*, Vol. 55, No. 2, pp. 251-276.

Frey, Giliola, and Matteo Manera. 2007. "Econometric Models of Asymmetric Price Transmission," *Journal of Economic Surveys*, Vol. 21, No. 2, pp. 349-415.

Goodwin, Barry, and Daniel Harper. 2000. "Price Transmission, Threshold Behavior, and Asymmetric Adjustment in the U.S. Pork Sector," *Journal of Agricultural and Applied Economics*, Vol. 32, No. 3, pp. 543-553.

Goodwin, Barry, and Matthew Holt. 1999. "Price Transmission and Asymmetric Adjustment in the U.S. Beef Sector," *American Journal of Agricultural Economics*, Vol. 81, pp. 630-637.

Hahn, William. 2004. *Beef and Pork Values and Price Spreads Explained*, LDP-M-188-01, USDA, Economic Research Service. Available at: http://usda.mannlib.cornell.edu/usda/ers/ldp-m//2000s/2004/ldp-m-05-10-2004_special_report.pdf/.

Hansen, Bruce. 1997. "Inference in TAR Models," *Studies in Nonlinear Dynamics and Econometrics*, Vol. 2, pp. 1-14.

Kejriwal, Mohitosh, and Pierre Perron. 2008. T*esting for Multiple Structural Changes in Cointegrated Regression Models*, Department of Economics, Boston University.

Kinnucan, Henry, and Olan Forker. 1987. "Asymmetry in Farm-Retail Price Transmission for Major Dairy Products," *American Journal of Agricultural Economics*, Vol. 69, No. 2, pp. 285-292.

MacDonald, James, and William McBride. 2009. *The Transformation of U.S. Livestock Agriculture: Scale, Efficiency, and Risks*, EIB-43, USDA, Economic Research Service. Available at: http://www.ers.usda.gov/publications/eib43/.

Martens, Martin, Paul Kofman, and Ton Vorst. 1998. "A Threshold Error-Correction Model for Intraday Futures and Index Returns," *Journal of Applied Econometrics*, Vol. 13, No. 3, pp. 245-263.

Meyer, Joche, and Stephan von Cramon-Taubadel. 2004. "Asymmetric Price Transmission: A Survey," *Journal of Agricultural Economics*, Vol. 55, No. 3, pp. 581-611.

Phillips, Peter, and Pierre Perron. 1988. "Testing for a Unit Root in Time Series Regression," *Biometrika*, Vol. 75, pp. 335-346.

Potter, Simon. 1995. "A Nonlinear Approach to US GNP," *Journal of Applied Econometrics*, Vol. 10, pp. 109-125.

Roeger, Edward, and Ephraim Leibtag. 2010. "The Magnitude and Timing of Retail Beef and Bread Price Response to Changes in Input Costs," presentation at Agricultural and Applied Economics Association Annual Meeting, July 25-27, 2010. Available at: http://purl.umn.edu/61041/.

Tsay, Ruey. 1989. "Testing and Modeling Threshold Autoregressive Processes," *Journal of the American Statistical Association*, Vol. 84, No. 405, pp. 231-240.

Vocke, Gary, Jean Buzby, and Hodan Farah Wells. 2009. "Consumer Preferences Change Wheat Flour Use," *Amber Waves*, Vol. 6, No. 4, p. 2.

APPENDIX: STATISTICAL TEST DESCRIPTIONS

Time-Series Properties of the Data

We began the investigation of the time series properties of the price data with unit root tests to establish the integration order of the individual series. This is necessary because in order for a cointegrating relationship to be possible, the series considered must be integrated of order 1—the series in levels is nonstationary but the first difference of the series is stationary. Cointegration implies that a stable, long-term relationship exists between time series variables that are themselves nonstationary. We used the modified Dickey-Fuller unit root test (DF-GLS) as described by Elliot et al., (1996) and included a trend term in the unit root test (if the time series data appeared to be following a linear trend over time). The results of the unit root tests, in all cases, failed to reject the presence of a unit root in the series itself, but rejected a unit root in the first difference of the series. Thus, the series were concluded to be first difference stationary.

The test for cointegration proceeds by looking at the stability of the long-term relationship between series by considering the stationarity of the residuals of the cointegrating relation. To begin, the cointegrating relation of equation 1 was estimated (by OLS) for each input series and its direct output series within a product group. The test for stationarity was then conducted on the residuals from the estimation with a Phillips-Perron unit root test, and this was done before any consideration of structural breaks. For all cases the null hypothesis of a unit root in the residuals, u, was rejected. This information combined with the result that the price series is integrated of order 1 suggests that a cointegrating relationship likely exists between the supply chain level price series.

Structural Breaks in the Longrun Equations

In order to estimate the number and timing of any structural changes in the cointegrated system, we used the procedure developed by Kejriwal and Perron (2008). Their 2008 work builds on an extensive literature that has systematically developed testing methods to determine the presence and placement of unknown change points within time series relationships (a detailed discussion of this progression can be found in this report's Introduction). Kejriwal and Perron look at several techniques for allowing different parts of the cointegrating equation to change over time (e.g. intercepts or all coefficients, although the statistical properties of the breaks become more difficult to determine if the parameter considered is nonstationary), and in our analysis the procedures are followed in which only the intercept is allowed to change.

We first consider the general test used by Kejriwal and Perron to confirm that some positive number of breaks may be appropriate. This is followed by using a sequential procedure that compares a model with k breaks against a model with $k+1$ breaks. The tests were conducted for each of the different price stage relationships following the form of equation 1 and with a 15-percent trimming rate (specifying that an area the size of 15 percent of the total observations would not be searched around a break), consistent with critical values calculated and presented by Kejriwal and Perron (2008). In each test, a maximum of three breaks is specified for this analysis to ensure that necessary sample sizes are maintained within each period. Details of the test statistics can be found in Roeger and Leibtag (2010).

In this analysis, we chose to include intercept shifts as the only time-varying parameters for a number of reasons, but this should not imply that future research should be limited to only this method. In our analysis we also considered allowing the coefficient on P_l to shift over time which would imply that, aside from the other input costs changing over time, the response to the main agricultural input is also time variant. In testing this alternative specification, where fi_l also shifts with the identified break dates, we found that this added little to the long-term models. This changed the residual values, u_l, in only minor ways and in comparing R^2 values, our initial adding of structural break dummy variable increased the amount of variance explained by the model by about 20 percent, while letting fi_l change across time as well, explained less than 1 percent of additional variance.

Threshold Bounds Search

The optimal bounds for the threshold ECMs, c_1 and c_2 (unique for each model), were found by conducting a grid search of the values of u. We use the method proposed by Balke and Fomby (1997) in which a grid search is conducted for threshold values that minimize the total sum of squared errors (SSE) across the conditional regression models. The restrictions on this search were that c_1 be greater than the lowest 15 percent of values, c_2 be less than the highest 15 percent, and a 15-percent band around 0 also be excluded. These 15-percent restrictions are utilized so that each regime has a sizeable enough number of observations to have its own separate regression model. The search proceeded by grouping together the u values using the c_1 and c_2 values as bounds on each group and estimating an autoregressive model for each group.

Following the restrictions above, different u values were sequentially used as bounds with the goal of the search to find the combination of c_1 and c_2 values that would produce the lowest total SSE from the estimation of the conditional autoregression models. The u values found for bounds c_1 and c_2 leading to the lowest total SSE, thus, provide the best grouping of negative and positive value cointegrating equation error terms, and allow the observations to be divided into three regimes based on where the particular u value falls for that observation.

Threshold Significance Test

In order to evaluate the statistical significance of threshold effects, we used the testing procedure introduced by Hansen (1997). In this test, a standard Chow test is performed and then repeated through a series of simulations using the same model but replacing the dependent variable values with a random draw in order to approximate the p-value for threshold significance. The test was performed for each threshold ECM with 350 repetitions. The null hypothesis of nonsignificance of the thresholds was rejected with an error rate of less than 0.001 percent for both wholesale to retail models, but at 19.4 percent for farm to wholesale beef and 17.7 percent for farm to whole-sale wheat flour. These results imply that for the retail models, the identified optimal threshold from the data are nearly always a better fit than groupings in random data, but for the wholesale models this is not the case 19.4 percent and 17.7 percent of the time (for beef and wheat flour, respectively).

Table A-1. Beef wholesale to retail pass-through regression estimation results

Model type	Variable	Coefficient		
Threshold ECM: Regime [1]	$\Delta(\ln$ beef PPI$)_{t-1}$	**	0.271	(0.043)
	Δ (ln beef PPI$)_{t-2}$	**	0.110	(0.055)
	Δ (ln grocery store wage$)_{t-2}$		0.035	(0.143)
	Δ (ln diesel$)_{t-6}$		0.002	(0.019)
	ECT$_{t-1}$	**	-0.120	(0.056)
Regime [2]	Δ (ln beef PPI$)_{t-1}$	**	0.218	(0.021)
	Δ (ln grocery store wage$)_{t-2}$		0.092	(0.080)
	Δ (ln diesel$)_{t-6}$		0.002	(0.009)
	ECT$_{t-1}$	**	0.071	(0.035)
Regime 3	Δ (ln beef PPI$)_{t-1}$	**	0.131	(0.029)
	Δ (ln beef PPI$)_{t-2}$	**	0.065	(0.027)
	Δ (ln grocery store wage$)_{t-2}$	*	0.133	(0.075)
Model type	Variable	Coefficient		
	Δ (ln diesel$)_{t-6}$		0.008	(0.009)
	ECT$_{t-1}$		-0.074	(0.047)
Symmetric	Δ (ln beef PPI$)_{t-1}$	**	0.247	(0.016)
	Δ (ln beef PPI$)_{t-2}$	**	0.065	(0.019)
	Δ (ln grocery store wage$)_{t-2}$	*	0.103	(0.057)
	Δ (ln diesel$)_{t-6}$		0.005	(0.007)
	ECT$_{t-1}$		-0.011	(0.011)

(*) denotes significance at least at the 10-percent level. (**) denotes significance at least at the 5-percent level.

Coefficients for constant and autoregressive terms are not listed.

The dependent variable in this regression model is the change in the downstream price: Δ(ln beef CPI$)_t$.

ECM = Error correction model; ECT = error correction term; PPI = Producer Price Index Source: USDA, Economic Research Service calculations based on U.S. Department of Labor, Bureau of Labor Statistics data.

Table A-2. Beef farm to wholesale pass-through regression estimation results

Model type	Variable	Coefficient		
Threshold ECM: Regime 1	$\Delta(\ln$ cattle PPI$)_{t-1}$	*	0.287	(0.150)
	Δ (ln slaugtering wage)$_{t-8}$		0.435	(0.507)
	Δ (ln diesel)$_{t-2}$		0.000	(0.054)
	ECT$_{t-1}$	*	-0.161	(0.096)
Regime 2	Δ (ln cattle PPI)$_{t-1}$	**	0.412	(0.085)
	Δ (ln slaugtering wage)$_{t-8}$	**	0.793	(0.262)
	Δ (ln diesel)$_{t-2}$		0.017	(0.022)
	ECT$_{t-1}$	**	-0.534	(0.183)
Regime 3	Δ (ln cattle PPI)$_{t-1}$	*	0.224	(0.119)
	Δ (ln slaugtering wage)$_{t-8}$		0.258	(0.315)
	Δ (ln diesel)$_{t-2}$	**	0.125	(0.044)
	ECT$_{t-1}$		-0.134	(0.090)
Symmetric ECM	Δ (ln cattle PPI)$_{t-1}$	**	0.340	(0.063)
	Δ (ln slaugtering wage)$_{t-8}$	**	0.600	(0.185)
	Δ (ln diesel)$_{t-2}$	**	0.038	(0.019)
	ECT$_{t-1}$	**	-0.136	(0.054)

(*) denotes significance at least at the 10-percent level.

(**) denotes significance at least at the 5-percent level.

Coefficients for constant and autoregressive terms are not listed.

The dependent variable in this regression model is the change in the downstream price: $\Delta(\ln$ beef PPI$)_t$.

ECM = Error correction model; ECT = error correction term; PPI = Producer Price Index

Source: USDA, Economic Research Service calculations based on U.S. Department of Labor, Bureau of Labor Statistics data.

Table A-3. Bread wholesale to retail pass-through regression estimation results

Model type	Variable	Coefficient		
Threshold ECM: Regime 1	$\Delta(\ln$ wheat f our PPI$)_{t-1}$	**	0.100	(0.033)
	$\Delta (\ln$ grocery store wage$)_{t-7}$		-0.060	(0.177)
	$\Delta (\ln$ diesel$)_{t-6}$		0.044	(0.037)
	ECT_{t-1}	**	-0.124	(0.054)
Regime 2	$\Delta (\ln$ wheat f our PPI$)_{t-3}$	**	0.052	(0.016)
	$\Delta (\ln$ wheat f our PPI$)_{t-4}$	**	0.056	(0.016)
	$\Delta (\ln$ grocery store wage$)_{t-7}$		0.078	(0.055)
	$\Delta (\ln$ diesel$)_{t-6}$		0.002	(0.006)
	ECT_{t-1}	**	-0.033	(0.013)
Regime 3	$\Delta (\ln$ wheat f our PPI$)_{t-2}$	**	0.062	(0.018)
	$\Delta (\ln$ wheat f our PPI$)_{t-3}$	**	0.046	(0.018)
	$\Delta (\ln$ wheat f our PPI$)_{t-4}$	**	0.047	(0.019)
	$\Delta (\ln$ grocery store wage$)_{t-7}$		0.052	(0.059)
	$\Delta (\ln$ diesel$)_{t-6}$		-0.004	(0.008)
	ECT_{t-1}		0.005	(0.022)
Symmetric	$\Delta (\ln$ wheat f our PPI$)_{t-1}$	**	0.051	(0.011)
	$\Delta (\ln$ wheat f our PPI$)_{t-2}$	**	0.037	(0.012)
	$\Delta (\ln$ wheat f our PPI$)_{t-3}$	**	0.058	(0.012)
	$\Delta (\ln$ wheat f our PPI$)_{t-4}$	**	0.039	(0.012)
	$\Delta (\ln$ grocery store wage$)_{t-7}$		0.038	(0.043)
	$\Delta (\ln$ diesel$)_{t-6}$		-0.002	(0.005)
	ECT_{t-1}	**	-0.009	(0.004)

(*) denotes significance at least at the 10-percent level.

(**) denotes significance at least at the 5-percent level.

Coefficients for constant and autoregressive terms are not listed.

The dependent variable in this regression model is the change in the downstream price: $\Delta(\ln$ white bread CPI$)_t$.

ECM = Error correction model; ECT = error correction term; PPI = Producer Price Index

Source: USDA, Economic Research Service calculations based on U.S. Department of Labor, Bureau of Labor Statistics data.

Table A-4. Bread farm to wholesale pass-through regression estimation results

Model type	Variable	Coefficient		
Threshold ECM: Regime 1	$\Delta(\ln$ wheat PPI$)_{t-1}$	**	0.112	(0.034)
	Δ (ln electricity price$)_{t-9}$	**	0.421	(0.210)
	Δ (ln diesel$)_{t-1}$		0.056	(0.058)
	ECT_{t-1}		-0.030	(0.047)
Regime 2	Δ (ln wheat PPI$)_{t-1}$	**	0.159	(0.085)
	Δ (ln wheat PPI$)_{t-2}$	**	0.104	(0.047)
	Δ (ln electricity price$)_{t-9}$		0.003	(0.136)
	Δ (ln diesel$)_{t-1}$	**	0.085	(0.036)
	ECT_{t-1}		-0.064	(0.132)
Regime 3	Δ (ln wheat PPI$)_{t-1}$	**	0.193	(0.074)
	Δ (ln electricity price$)_{t-9}$	**	0.395	(0.128)
	Δ (ln diesel$)_{t-1}$		-0.012	(0.028)
	ECT_{t-1}	**	-0.090	(0.045)
Symmetric	Δ (ln wheat PPI$)_{t-1}$	**	0.115	(0.047)
	Δ (ln electricity price$)_{t-9}$	*	0.162	(0.088)
	Δ (ln diesel$)_{t-1}$	*	0.042	(0.022)
	ECT_{t-1}	*	-0.069	(0.036)

(*) denotes significance at least at the 10-percent level. (**) denotes significance at least at the 5-percent level.

Coefficients for constant and autoregressive terms are not listed.

The dependent variable in this regression model is the change in the downstream price: Δ (ln wheat fl our PPI$)_t$.

ECM = Error correction model; ECT = error correction term; PPI = Producer Price Index

Source: USDA, Economic Research Service calculations based on U.S. Department of Labor, Bureau of Labor Statistics data.

ACKNOWLEDGMENTS

The authors thank Barry Goodwin, William Neal Reynolds Distinguished Professor, Department of Agricultural and Resource Economics, North Carolina State University– Raleigh; Donald Liu, Professor, Department of Applied Economics, University of Minnesota; and ERS colleague William Hahn for their reviews, Priscilla Smith for editorial assistance, and Wynnice Pointer-Napper for design assistance.

End Notes

[1] See the ERS web briefing room, Food CPI and Expenditures, for more information about the forecasts, at: http://www.ers.usda.gov/briefi ng/cpifoodandexpenditures/.

[2] In the case of the farm-to-wholesale beef model, the sample starts with 1976 due to data limitations.

[3] For example, for retail beef, the overall Consumer Price Index (CPI) value for beef and veal is used instead of pricing information for a specific beef product. This avoids the problems associated with having to track back the production origins of a specific product that may not be representative of the larger category.

[4] The conversion to natural logs allows for the interpretation of the estimated pass-through coefficients to be in terms of proportional price movements.

[5] For both the slaughtering wage and grocery-store wage, we use data from the Bureau of Labor Statistics National Compensation Survey.

[6] In comparison, the rise in the retail bread CPI from the beginning of 1997 to the end of 2002 was slightly less than 18 percent.

[7] A summary of USDA, ERS meat price-spread data are available at http://www.ers.usda.gov/data/meatpricespreads/. Field crops price-spread data are available at http://www.ers.usda.gov/ data/farmtoconsumer/pricespreads.htm/.

[8] While structural break dates were identified at specific points, they may represent shifts that take place over longer periods of time, as well. More detail on the tests to determine the number and placement of the structural break dates is given in the appendix.

[9] The production locus represents the number at which 50 percent of the cattle operations were smaller than this number and 50 percent were larger.

[10] Contemporaneous impacts on the dependent variable from the exog- enous variables were not considered in this analysis. The number of lags we include for each variable in the model was determined by using the Hannan- Quinn information criterion.

[11] As discussed in more detail in the next section, a threshold model allows for the response to changes in input prices to differ depending on the magni- tude and sign of the input price change.

[12] Again, these lag lengths were chosen using the Hannan-Quinn information criterion.

[13] See appendix for details on how c1 and c2 are determined.

[14] This test is described in Hansen (1997); more details on the procedure that we used are given in this report's appendix.

[15] Estimates of the autoregressive terms in the models are not included in the regression output tables, though these terms were included when estimating the models.

[16] The retail beef result given here is a range of numbers since the threshold model results have three different sets of coeff cients (one for each regime). The wholesale beef result is only one coeff cient since the symmetric ECM has one set of results for each product and price relationship.

[17] In the threshold ECMs, the sample splitting makes the interpretation of the estimates of these variables more difficult.

[18] The threshold model is sensitive to the time period in question because the threshold bounds are in terms of a variable that is sensitive to the deviations in the current relationship from the expected long-term relationship of PO and PI.

[19] This method is described in further detail in Potter (1995).

[20] The response functions are based on impulses in which all input price changes (in percent terms) are the same as the actual input price changes after the date of the impulse, so that our estimates isolate the impact of a one-time change.

[21] The NLIRF analysis was sequentially repeated over this time period because the threshold models (and consequently the NLIRF) are sensitive to time-period considerations and our goal was to present an average response.

[22] See, for example, Abdulai (2002) and Goodwin and Holt (1999).

In: Consumer Food Costs
Editor: Stefanee L. Martin

ISBN: 978-1-61470-695-3
© 2011 Nova Science Publishers, Inc.

Chapter 3

HOW MUCH DO FRUITS AND VEGETABLES COST?[*]

Hayden Stewart, Jeffrey Hyman, Jean C. Buzby, Elizabeth Frazão and Andrea Carlson

ABSTRACT

Federal dietary guidance advises Americans to consume more vegetables and fruits because most Americans do not consume the recommended quantities or variety. Food prices, along with taste, convenience, income, and awareness of the link between diet and health, shape food choices. We used 2008 Nielsen Homescan data to estimate the average price at retail stores of a pound and an edible cup equivalent (or, for juices, a pint and an edible cup equivalent) of 153 commonly consumed fresh and processed fruits and vegetables. We found that average prices ranged from less than 20 cents per edible cup equivalent to more than $2 per edible cup equivalent. We also found that, in 2008, an adult on a 2,000- calorie diet could satisfy recommendations for vegetable and fruit consumption in the *2010 Dietary Guidelines for Americans* (amounts and variety) at an average price of $2 to $2.50 per day, or approximately 50 cents per edible cup equivalent.

---

[*] This is an edited, reformatted and augmented version of the United States Department of Agriculture publication Economic Research Service, Economic Information Bulletin 71, dated February 2011.

Keywords: food prices, food budgeting, fruit and vegetable consumption, *2010 Dietary Guidelines for Americans.*

SUMMARY

What Is the Issue?

Federal dietary guidance advises Americans to consume more vegetables and fruits because most Americans do not consume the recommended quantities or variety. Food prices, along with taste, convenience, income, and awareness of the link between diet and health, shape food choices. This research updates previous estimates of vegetable and fruit prices, and estimates the cost of satisfying recommendations for adult vegetable and fruit consumption in the 2010 Dietary Guidelines for Americans.

What Are the Major Findings?

We estimated the average retail prices of 153 fresh and processed vegetables and fruits, where processed includes frozen, canned, and dried vegetables and fruits as well as 100% fruit juice. We also estimated the average price per edible cup equivalent for each vegetable and fruit. This is the consumption unit used in the 2010 Dietary Guidelines for Americans, and measures only the edible portion of a food once it has been cooked or otherwise prepared for consumption. In 2008:

- An adult on a 2,000-calorie diet could satisfy recommendations for vegetable and fruit consumption (amounts and variety) in the *2010 Dietary Guidelines for Americans* at an average cost of $2 to $2.50 per day, or approximately 50 cents per edible cup equivalent.
- The lowest average price for any of the 59 fresh and processed fruits included in the study was for fresh watermelon, at 17 cents per edible cup equivalent. The highest average price was for fresh raspberries, at $2.06 per edible cup equivalent.
- The lowest average price for any of the 94 fresh and processed vegetables included in the study was for dry pinto beans, at 13 cents per edible cup equivalent. The highest average price was for frozen asparagus cuts and tips, at $2.07 per edible cup equivalent.

- Processed fruits and vegetables were not consistently more or less expensive than fresh produce. Canned carrots (34 cents per edible cup equivalent) were more expensive than whole fresh carrots eaten raw (25 cents per edible cup equivalent). However, canned peaches (58 cents per edible cup equivalent) were less expensive than fresh (66 cents per edible cup equivalent).
- Retail prices per pound often varied substantially from prices per edible cup equivalent. Fresh broccoli florets and fresh ears of sweet corn both sold for around $1.80 per pound at retail stores, on average. After boiling and removing inedible parts, however, the sweet corn cost almost twice as much as the broccoli florets ($1.17 vs. 63 cents per edible cup equivalent).

Costs in the study are defined as the average prices paid by all American households for a food over a 1-year period, including purchases in different package sizes, under different brand names, and at different types of retail outlets (including, among others, supercenters such as Wal-Mart, wholesale club stores such as Costco, "traditional" grocers such as Safeway, Kroger, and Albertsons, and convenience stores).

How Was the Study Conducted?

We used 2008 Nielsen Homescan data to calculate the average price of a pound (or, for juices, a pint) of 153 fresh and processed fruits and vegetables at retail stores. In order to estimate price per edible cup equivalent for each food, retail quantities were adjusted for the removal of inedible parts and cooking that occur prior to consumption. For example, 1 pound of store-bought fresh pineapple yields 0.51 pound of edible pineapple. Data from the USDA National Nutrient Database for Standard Reference (Release 21) and USDA's *Food Yields Summarized by Different Stages of Preparation* were used to estimate edible weights. The MyPyramid Equivalents Database, 2.0 was used to define edible cup equivalents.

INTRODUCTION

Federal dietary guidance advises Americans to consume more vegetables and fruits because most Americans do not consume the recommended quantities or variety. Individuals choose foods based on taste, convenience, income, and awareness of the link between diet and health. Food prices also shape food choices.

How much do fruits and vegetables cost? Reed et al. (2004) used 1999 Nielsen Homescan data to estimate average prices at retail stores across the contiguous United States. They found that consumers could purchase enough fruits and vegetables to satisfy Federal dietary recommendations in place between 2000 and 2004 for $1 a day (three servings of fruits and four servings of vegetables). The USDA Center for Nutrition Policy and Promotion (CNPP) also provides estimates of average prices for fruits and vegetables. As of December 2010, the online CNPP Prices Database reported average prices paid per 100 edible grams in 2003-04. Here, we use 2008 data on prices to update these previous estimates of the cost of vegetables and fruits.

We also use the newly released *2010 Dietary Guidelines for Americans* to estimate the cost of satisfying recommended adult vegetable and fruit consumption. The U.S. Department of Agriculture and U.S. Department of Health and Human Services revise the Dietary Guidelines every 5 years to provide up-to-date authoritative advice for people 2 years old and older about how good dietary habits can promote health and reduce risk for major chronic diseases. The Guidelines also serve as the basis for Federal food and nutrition education programs. In 2005, the Dietary Guidelines increased recommended intakes of vegetables and fruits for most Americans and made more specific recommendations about how to divide vegetable consumption among subgroups (Guenther et al., 2006).

According to the *2010 Dietary Guidelines for Americans*, a person on a 2,000-calorie diet needs 2.5 cup equivalents of vegetables per day or, equivalently, 17.5 cup equivalents per week (table 1). Weekly consumption ought to include dark green vegetables (1.5 cup equivalents), red and orange vegetables (5.5 cup equivalents), beans and peas (1.5 cup equivalents), starchy vegetables (5 cup equivalents), and other vegetables (4 cup equivalents).[1] Fresh, canned, frozen, dried, and 100% juice count equally toward recommended intakes, although the majority of the recommended fruit should come from whole fruit.

Table 1. Dietary recommendations tailored to gender, age, and level of physical activity

Consumer (gender and age)[1]	Total calories	Fruit cup equivalents	Vegetable cup equivalents
	Daily		
Female, 6 years old	1,400	1.5	1.5
Male, 6 years old	1,600	1.5	2
Female, 20 years old	2,200	2	3
Female, 40 years old	2,000	2	2.5
Male, 20 years old	2,800	2.5	3.5
Male, 40 years old	2,600	2	3.5

[1] All people are assumed to be moderately active (30-60 minutes of exercise daily).
Source: *2010 Dietary Guidelines for Americans*.

HOW DO WE ESTIMATE THE COST OF FRUITS AND VEGETABLES?

We estimate average retail prices using 2008 Nielsen Homescan data with information on 61,440 households. Households participating in Nielsen's Homescan panel keep a scanner in their home to record their purchases of foods at retail stores. After a shopping occasion, panelists use these scanners to record the items purchased, the quantities bought, the amount of money paid, and the date. Purchases at supercenters (such as Wal-Mart), wholesale club stores (such as Costco), traditional grocery stores, convenience stores, drugstores, and other types of retail facilities are all included. Nielsen further provides sample weights that allow data users to estimate what all households across the contiguous U.S. paid for foods at retail stores and the quantities they bought.[2]

We define the cost of each food as the average price paid by all American households for that food over a 1-year period, including purchases in a variety of package sizes, under different brand names, and at several types of retail outlets. Costs to buy foods at retail stores are estimated on a per-pound (or, for 100% juices, a per-pint) basis, while the costs to consume foods are estimated on a per-edible-cup-equivalent basis. Further details on how we applied this methodology to specific fruits and vegetables are available on our website, at: http://www.ers.usda.gov/data/FruitVegetableCosts/index.htm/.

Selecting the Foods to Price

Many types of fruits and vegetables are available at retail stores across the Nation. For this study, we focused on fruits and vegetables that account for a large share of total consumption. Some less commonly consumed types, such as guava, are excluded, as are lemons, limes, garlic, and other types used as flavorings and condiments but not widely consumed alone as a food.

Selected types of fruits and vegetables are priced in various fresh and processed forms, where processed includes frozen, canned, and dried vegetables and fruits as well as 100% fruit juice. For example, we priced fresh apples, dried apples, and unsweetened applesauce. Apples are also priced in two 100% juice forms— ready-to-drink and frozen concentrate that must be reconstituted at home.

The last step in our selection process was to identify fairly specific products for pricing. Data constraints influenced our selection among fresh fruit and vegetable products at this step. Specifically, we had to exclude fresh produce sold on a "random-weight" basis, such as whole, untrimmed heads of Romaine lettuce. Marketers usually sell untrimmed heads of Romaine lettuce in loose form. Consumers can choose among the heads on display and place their selection in a plastic bag. The weight of the food placed in the bag is not fixed. Thus, in retail terminology, the lettuce is sold on a random-weight basis. Nielsen did not provide data on sales of individual random-weight foods in 2008. Because of this, we could not price heads of Romaine lettuce and other random-weight produce, but we could price products like Romaine hearts, which are generally sold in bags that include a manufacturer's or retailer's brand name along with a Universal Product Code (UPC, a type of bar code).

How do prices compare for produce sold on a random-weight basis and produce marketed in other ways? To investigate, we used 2006 Nielsen Homescan data. When Homescan panelists bought fresh produce on a random-weight basis in 2006, Nielsen identified the specific type of fruit or vegetable bought. Prices for random-weight produce can then be compared with prices for fruits and vegetables marketed in other ways including food products sold in a prepackaged container or on a "count" basis, such as a $2.50 per melon (table 2).

For some types of fruits and vegetables, such as whole carrots and celery stalks, there is very little difference in price between produce sold on a random-weight basis and the same type of produce sold in a package with a manufacturer's or retailer's brand name. For example, using 2006 price data,

How Much Do Fruits and Vegetables Cost? 111

we estimate that random-weight whole carrots cost about 69 cents per pound while prebagged whole carrots cost 67 cents per pound.

Substantial price differences do exist for other types of produce depending on how the food is marketed. In 2006, apples sold on a random-weight basis ($1.18 per pound) cost about 30 cents more per pound than apples sold in prepackaged form (85 cents per pound). One possible explanation is that more varieties are sold on a random-weight basis, some of which may be more expensive than the varieties that are available in bagged forms. Another possibility is that retailers prefer to sell the largest and most attractive apples in loose form, leaving smaller and less attractive apples for bagging.

Table 2. Retail prices for random weight and other fresh produce, 2006 prices[1]

	Random weight[2]	Other fresh produce[3]
		Dollars per pound
Apples	1.18	0.85
Broccoli heads	1.07	--
Broccoli florets	--	1.79
Celery stalks	0.85	0.82
Carrots, whole	0.69	0.67
Carrots, baby	--	1.36
Oranges (all varieties)	0.83	0.61
Potatoes	0.67	0.41
Romaine heads	1.19	--
Romaine hearts	--	1.74
Spinach, bunch	1.05	--
Spinach, leaf and baby	--	3.32

[1] Average retail prices per pound, not adjusted for the removal of inedible parts and cooking that may be required prior to consumption.

[2] Random weight produce includes loose items from which consumers can choose the quantity of products they want, generally by placing their desired fruit or vegetable in a plastic bag. [3]Includes items sold in a prepackaged container, such as a bag or clamshell, and items sold on a count basis, such as oranges priced per piece of fruit.

-- = not available or insufficient data accessible.

Source: USDA, Economic Research Service analysis of 2006 Nielsen Homescan data.

The largest cost differentials exist for higher value-added fresh fruits and vegetables. Many processors are adding more value to produce by removing inedible parts and/or washing it. They also tend to sell these foods in a bag or clamshell that includes either their name or a retailer's brand name. Bagged broccoli florets are one such product. In 2006, this product cost $1.79 per pound whereas random-weight heads of broccoli cost $1.07. However, broccoli florets include no refuse, whereas households may discard much of the stem on a head of broccoli. Because of this, using 1999 data, Reed et al. (2004) found that florets can be cheaper to consume per serving than heads of broccoli.

In general, however, higher value-added produce costs more money to consume than traditional produce even after accounting for food parts that are discarded. For example, a person wanting to consume fresh spinach could purchase a random-weight bunch at a lower price than our estimated price for leaf and baby spinach sold in prepackaged bags or clamshells. In 2006, prepackaged leaf and baby spinach cost $3.32 per pound while random-weight bunch spinach cost $1.05 per pound. Even though households may discard the stems and roots on a bunch of spinach, it is likely that they can still economize by purchasing this product over the higher value-added product in a bag or clamshell. For 2008, along with fresh spinach, we could only price fresh broccoli, Romaine lettuce, mustard greens, turnip greens, collard greens, kale, and winter squash in higher value-added forms.

Overall, retail prices for random-weight fruits and vegetables are neither higher nor lower than prices for produce marketed in other ways. For our analysis of 2008 food prices, it follows that households could have saved money by purchasing random-weight produce instead of the food products in our sample, in cases where the random-weight items were available at a lower price. By contrast, in cases where the random-weight produce was more expensive, economizing households could have still bought the foods in our sample.

The processed foods included in our study are somewhat similar to their fresh counterparts. Our intention was to make the fresh and processed foods in our study as comparable as possible to each other in nutritional quality. For that reason, we excluded apple juice blended with other juices and banana chips made with oil. We did include sweetened and flavored foods in some cases because excluding all sweetened or flavored foods would have overly restricted our sample. Thus, we included canned peaches packed in syrup.

In total, we priced 153 food products including fresh, canned, frozen, dried, and juiced items (tables 3 and 4). We further classified each of these

How Much Do Fruits and Vegetables Cost? 113

food products as either a fruit or vegetable according to how Federal dietary recommendations classify the same food. Avocadoes, mushrooms, olives, and tomatoes were classified as vegetables because the consumption of these foods counts toward an individual's recommended intake of vegetables.

Table 3. Fruits included in the study

Apples	Figs, dried	Peaches
Fresh	Grapefruit	Fresh
Canned, applesauce	Fresh	Canned
Juice	Canned	Pears
Ready to drink	Juice	Fresh
Frozen	Ready to drink	Canned
Dried	Frozen	Pineapple
Apricots	Grapes	Fresh
Canned	Fresh	Canned
Dried	Juice	Juice
Bananas, fresh	Ready to drink	Ready to drink
Blackberries	Frozen	Frozen
Fresh	Dried, raisins	Dried
Canned	Honeydew, fresh	Plums/prunes
Frozen	Kiwi, fresh	Fresh
Blueberries	Mangoes	Juice, ready to drink
Fresh	Fresh	Dried prunes
Canned	Dried	Raspberries
Frozen	Nectarines, fresh	Fresh
Cantaloupe, fresh	Oranges	Frozen
Cherries	Fresh navel	Strawberries
Fresh	Canned Mandarin	Fresh
Canned	Juice	Canned
Sweet	Ready to drink	Frozen
Tart	Frozen	Tangerines
Cranberries, dried	Papayas, fresh	Fresh
Dates, dried		Juice, ready to drink
Watermelon, fresh		
Apples	Figs, dried	Apples
Total fruit		
Fresh	22	
Canned	12	
Frozen	4	
Juiced	12	
Dried	9	
All fruit	59	

Source: USDA, Economic Research Service.

Table 4. Vegetables, beans, and peas included in the study

Artichoke	Cauliflower	Lentils, dried	Pumpkin, canned
Fresh	Fresh	Lima beans	Radishes, fresh
Canned	Florets	Canned	Red kidney beans
Frozen	Heads	Frozen	Dried
Asparagus	Frozen	Dried	Canned
Fresh spears	Celery, fresh	Mushrooms	Red peppers, bell fresh
Canned	Hearts	Fresh	Romaine hearts, fresh
Cut & tips	Stalks	Sliced	Spinach
Spears	Collard greens	Whole	Fresh
Frozen	Fresh	Canned	Canned
Cut & tips	Canned	Frozen	Frozen
Spears	Frozen	Mustard greens	Squash, summer
Avocados, fresh	Corn, sweet	Fresh	Fresh
Beets, canned	Corn on cob, fresh	Canned	Canned
Black beans	Whole kernel	Frozen	Frozen
Dried	Canned	Navy beans	Squash, winter
Canned	Frozen	Dried	Fresh
Broccoli	Great Northern beans	Canned	Frozen
Fresh, florets	Dried	Okra	Sweet potatoes
Frozen	Canned	Fresh	Fresh
Brussels sprouts	Green beans	Canned	Frozen, french fries
Fresh	Fresh	Frozen	Tomatoes
Frozen	Canned	Olives, canned	Fresh
Cabbage	Cut	Onions, fresh	Cherry & grape
Fresh	Whole	Peas, green	Roma & plum
Canned sauerkraut	Frozen	Canned	Round
Carrots	Cut	Frozen	Canned
Fresh	Whole	Pinto beans	Turnip greens
Baby	Green peppers, bell fresh	Dried	Fresh
Whole	Iceberg lettuce, fresh	Canned	Canned
Canned	Kale	Potatoes, white	Frozen
Sliced	Fresh	Fresh	
Whole	Canned	Canned	
Frozen	Frozen	Frozen, french fries	
Total vegetables			
Fresh	35		
Canned	24		
Frozen	23		
Juiced	0		
Beans and peas	12		
All vegetables, beans, and peas	94		

Source: USDA, Economic Research Service.

Estimating the Average Retail Prices of Selected Foods

The average retail price of each food was estimated on a per-pound (or, for juices, a per-pint) basis. To do so, we first used the 2008 Homescan data to estimate total expenditures by U.S. households on each food and the total quantities bought. Average retail prices then were calculated as the ratio of total expenditures to total quantities. We estimated that Americans spent $247.1 million on frozen concentrated orange juice, which, if reconstituted, could make 480.7 million pints. Thus, the average retail price of frozen concentrated juice was estimated at 51 cents per pint ($247.1 million/480.7 million pints).

To estimate total expenditures and quantities, we aggregated over purchases made by all households, in all seasons of the year, in all package sizes, and at all retail store formats. We also used Nielsen's sample weights to make our estimates representative of what all households across the contiguous United States paid in 2008.

While calculating aggregate household expenditures on each type of food was straightforward, calculating aggregate quantities of foods purchased by households was more complicated. Fruits and vegetables are sold primarily by the pound or ounce. For example, whole fresh carrots are typically sold in bags weighing 1, 2, or 5 pounds. However, we had to impute a weight for products sold on a count basis, such as melons or oranges priced per piece of fruit. To convert these sales to a weight basis, we used the USDA National Nutrient Database for Standard Reference, Release 21 (SR). The USDA Standard Reference estimates the weight of a medium cantaloupe at about 1,082 grams (roughly 2.4 pounds), including the weight of the rind and inedible cavity contents such as seeds.

Estimating Average Prices per Edible Cup Equivalent

We also estimated the average price of each fruit and vegetable per edible cup equivalent as defined in the *MyPyramid Equivalents Database*, Version 2.0 (MPED). For many fruits and vegetables, a 1-cup equivalent equals the weight of enough edible food after cooking, if necessary, to fill a measuring cup. For example, a cup equivalent of cooked whole kernel corn weighs 164 grams whether from fresh, frozen, or canned product. However, there are exceptions. To make a 1-cup equivalent, it takes 2 edible cups of a raw, leafy vegetable, like spinach, but only one-half cup of edible dried fruit.[3] Prior to

2005, USDA provided nutritional advice in servings. A serving of a fruit or vegetable generally equals one-half cup.

Our estimates of the costs to consume fruits and vegetables are based on an approach developed by USDA's Center for Nutrition Policy and Promotion (Carlson et al., 2008). We also rely on USDA's Standard Reference and *Food Yields Summarized by Different Stages of Preparation* (Handbook 102) for data on cooking yields and data on the edible shares of fruits and vegetables. If weight is lost through preparation, we define a food's retail-equivalent weight as:

Retail-equivalent weight = weight of a cup equivalent/(1 - share lost)

where shares are expressed as fractions. For example, the Standard Reference (SR) reports that 10 percent of a fresh apple is inedible, while the MPED lists the weight of a 1-cup equivalent of raw apple with skin at 106 grams. To eat a 1-cup equivalent, households must therefore buy 106/0.9 = 117.78 grams of whole fresh apple. By contrast, if weight is gained through preparation, we define a food's retail-equivalent weight as:

Retail-equivalent weight = weight of a cup equivalent/(1 + share gained)

where shares are again expressed as fractions. USDA Handbook 102 reports that cooking dry beans increases their weight. The weight of the cooked product is approximately 240 percent of the weight of the dry beans prior to cooking. The MPED further lists the weight of a 1-cup equivalent of cooked pinto beans at 173 grams. Households must therefore buy 173/2.4 = 72.08 grams of dry pinto beans at a retail store to eat a 1-cup equivalent at home.

Finally, because cup equivalent weights are in grams, we converted our earlier estimates of retail prices from a dollars-per-pound basis to a dollarsper-gram basis (by dividing by 453.59), and calculated the cost to eat a cup equivalent of a food as:

Price per cup equivalent = (average retail price per gram) x
(retail-equivalent weight in grams).

Having estimated the average costs to buy and to consume selected fruits and vegetables, we report our results in figures 1-9. As noted, our sample of foods does not include all fruit and vegetable products available at retail stores. There are still other products available at both higher and lower prices than the 153 foods we examined.

AVERAGE FRUIT PRICES IN 2008

We priced 59 fruit products: 22 fresh, 12 canned, 4 frozen, 12 juiced, and 9 dried (see table 3). Average retail prices for selected fruit in 2008 ranged from 26 cents per pound to over $7 per pound. After adjusting retail food prices for the inedible parts of fruit products and cooking yields, there was still much variability on a cup equivalent basis. Below, we examine average fruit prices.

Fresh Fruit

How much does fresh fruit cost? Out of the 22 types of fresh fruit we analyzed, retail prices ranged from 26 cents per pound for watermelon to $7.29 per pound for raspberries (fig. 1a). Two of these fruits (bananas and watermelon) cost less than 50 cents per pound, 6 cost less than $1 per pound, and 10 cost between $1 and $2 per pound.

How much does it cost to eat a cup equivalent of fresh fruit? We assume that all fresh fruits are consumed fresh. Only inedible parts such as the stem and core of an apple or peel and navel of a navel orange are removed prior to consumption and discarded. Under these assumptions, eight fresh fruits cost less than 50 cents per cup equivalent (fig. 1b). Watermelon was the least expensive at 17 cents per cup equivalent followed by bananas (21 cents), apples (28 cents), and navel oranges (34 cents).

Fresh berries, papayas, and cherries were the most expensive types of fresh fruits to eat. Fresh raspberries cost $2.06 per cup equivalent, on average. Among fresh berries and cherries, we estimate that strawberries were least expensive, at 89 cents per cup equivalent.

Canned Fruit

Of the 12 canned fruits in the study, applesauce was the least expensive to buy at $0.85 per pound (fig. 2a). Canned pineapple, Mandarin oranges, peaches, and pears all sold for between 90 cents and $1.10 per pound, on average.

Canned fruits may be used as an ingredient in baked or frozen foods or simply eaten without further preparation. For this study, we use the weight for a cup equivalent reported in the MPED for canned fruits that are eaten without further preparation. Using this approach, the cost to eat canned fruits ranged from 46 cents to $1.60 per cup equivalent (fig. 2b). Applesauce and pineapple cost less than 50 cents per cup equivalent. Mandarin oranges, pears, and peaches cost less than 60 cents per cup equivalent. Canned blueberries and blackberries were the most expensive canned fruits to consume, at more than $1.50 per cup equivalent.

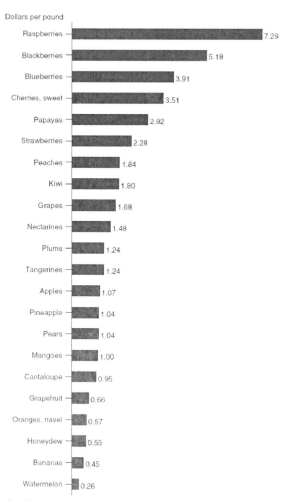

Note: Prices are for fruit sold in a prepackaged container, such as in a bag or clamshell, and fruit sold on a count basis, such as melons and oranges sold per piece of fruit.
Source: USDA, Economic Research Service analysis of 2008 Nielsen Homescan data.
Figure 1a. Fresh fruit: Average retail prices.

How Much Do Fruits and Vegetables Cost?

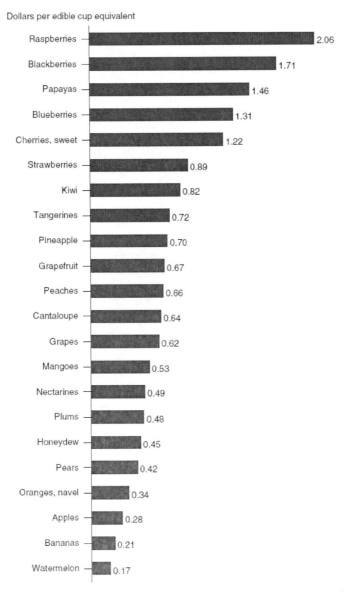

Note: Prices are for fruit sold in a prepackaged container, such as in a bag or clamshell, and fruit sold on a count basis, such as melons and oranges sold per piece of fruit. Edible cup equivalents are units of measurement for fruit/vegetable-consumption recommendations.

Source: USDA, Economic Research Service analysis of 2008 Nielsen Homescan data.

Figure 1b. Fresh fruit: Average prices per edible cup equivalent .

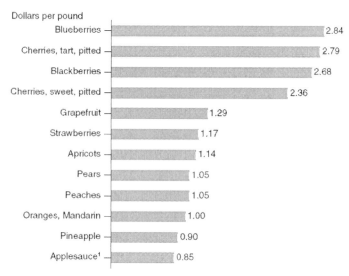

[1] Includes unsweetened and "diet" varieties.
Source: USDA, Economic Research Service analysis of 2008 Nielsen Homescan data.
Figure 2a. Canned fruit: Average retail prices.

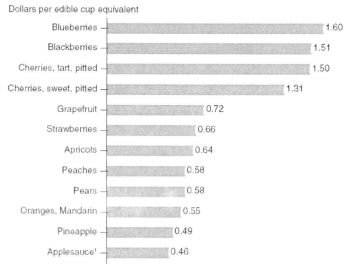

Note: Edible cup equivalents are units of measurement for fruit/vegetable-consumption recommendations. 1Includes unsweetened and "diet" varieties.
Source: USDA, Economic Research Service analysis of 2008 Nielsen Homescan data.
Figure 2b. Canned fruit: Average prices per edible cup equivalent.

Frozen Fruit

Four popular types of frozen berries, all unsweetened, are included in our food cost analysis. Strawberries were least expensive costing $2.12 per pound, on average (fig. 3a). Blueberries, blackberries, and raspberries all cost over $3 per pound.

We assume that frozen berries include only edible fruit. All inedible parts are removed prior to packaging for retail sale. We further assume that frozen raspberries, blueberries, and blackberries are eaten frozen, such as in a fruit smoothie or other dessert. However, because the MPED does not provide a weight for frozen strawberries, we must assume that frozen strawberries are thawed prior to consumption. Under these assumptions, frozen berries cost $1 to $2 per cup equivalent, depending on the type of fruit (fig. 3b).

However, thawed strawberries ($1.14 per cup equivalent) were not cheaper to eat than were frozen blackberries ($1.13 per cup equivalent). Frozen raspberries were the most expensive frozen berry to eat at $1.86 per cup equivalent, on average.

Fruit Juice

We priced 12 100% juice products, including several types of fruit juice in both frozen concentrated and ready-to-drink (shelf stable or refrigerated) forms. Frozen concentrated apple juice was the least expensive product to buy at 40 cents per reconstituted pint and the least expensive to drink at 20 cents per reconstituted cup equivalent (figs. 4a and 4b). Many other varieties of juice were also available for less than 30 cents per cup equivalent including frozen concentrated grape, orange, pineapple, and grapefruit juice. Only ready-to-drink tangerine and prune juice cost more than 50 cents per cup equivalent. Generally speaking, frozen concentrated juices cost less per pint and per cup equivalent than ready-to-drink juices.

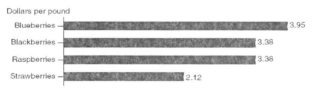

Note: Includes only unsweetened, frozen fruits.
Source: USDA, Economic Research Service analysis of 2008 Nielsen Homescan data.
Figure 3a. Frozen fruit: Average retail prices.

Note: Includes only unsweetened, frozen fruits. Edible cup equivalents are units of measure-ment for fruit/vegetable-consumption recommendations.

Source: USDA, Economic Research Service analysis of 2008 Nielsen Homescan data.

Figure 3b. Frozen fruit: Average prices per edible cup equivalent.

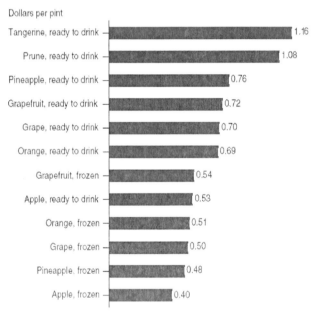

Note: "Ready to drink" refers to the state of the juice at the point of purchase. This includes juice reconstituted by the manufacturer from concentrate, as well as juice not from concentrate."Frozen" refers to juice sold as frozen concentrate, which consumers can dilute with water at home.

Source: USDA, Economic Research Service analysis of 2008 Nielsen Homescan data.

Figure 4a. Fruit juice: Average retail prices.

How Much Do Fruits and Vegetables Cost?

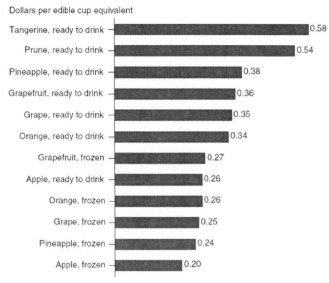

Note: "Ready to drink" refers to the state of the juice at the point of purchase. This includes juice reconstituted by the manufacturer from concentrate, as well as juice not from concentrate. "Frozen" refers to juice sold as frozen concentrate, which consumers can dilute with water at home. Edible cup equivalents are units of measurement for fruit/vegetable-consumption recommendations.

Source: USDA, Economic Research Service analysis of 2008 Nielsen Homescan data.

Figure 4b. Fruit juice: Average prices per edible cup equivalent.

Dried Fruit

We priced nine types of dried fruit for our food cost analysis. Raisins were the least expensive, costing $2.42 per pound, on average (fig. 5a). Dried mango, apples, figs, and pineapple all cost over $4 per pound.

We assume that dried fruit products contain no inedible parts and cooking is not required. The fruits are sold ready-to-eat. Under these assumptions, we found that raisins were the least expensive dried fruit to consume at 39 cents per cup equivalent (fig. 5b). Apricots, cranberries, apples, and prunes cost a few cents more than 50 cents per cup equivalent. Figs ($1.08/cup equivalent), mango (79 cents/cup equivalent), and pineapple (70 cents/cup equivalent) were the most expensive dried fruit to consume, on average.

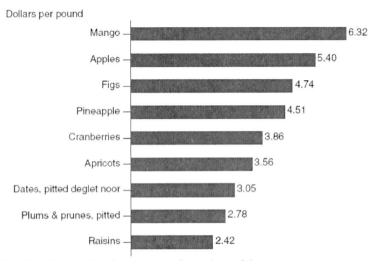

Note: "Pitted deglet noor" is the most popular variety of date.
Source: USDA, Economic Research Service analysis of 2008 Nielsen Homescan data.
Figure 5a. Dried fruit: Average retail prices.

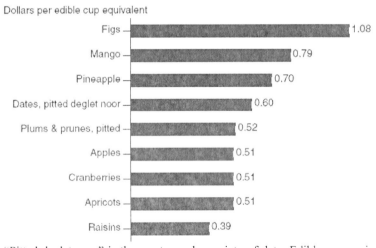

Note: "Pitted deglet noor" is the most popular variety of date. Edible cup equivalents are the units of measurement for fruit/vegetable-consumption recommendations.
Source: USDA, Economic Research Service analysis of 2008 Nielsen Homescan data.
Figure 5b. Dried fruit: Average prices per edible cup equivalent.

Average Vegetable Prices in 2008

We priced 94 vegetable products: 35 fresh, 24 canned, 23 frozen, and 12 beans and peas (see table 4). Average retail prices for vegetables ranged from less than 50 cents per pound to over $5 per pound. After adjusting retail food prices for inedible shares and cooking yields, there was still much variability in average prices on a cup equivalent basis. Below, we examine average vegetable prices, including beans and peas.

Fresh Vegetables

How much do fresh vegetables cost at retail? Of the 35 fresh products included in our study, 8 cost less than $1 per pound. These include potatoes, cabbage, onions, heads of cauliflower, whole carrots, celery stalks, sweet potatoes, and heads of iceberg lettuce (fig. 6a). The most expensive products were sliced mushrooms ($4.02 per pound) and fresh-cut leaf and baby spinach ($3.92 per pound).

Many vegetables that are bought fresh also are eaten raw. Other vegetables, such as potatoes, must be boiled, steamed, or otherwise cooked. Still others, such as carrots, are widely eaten both raw and cooked. Thus, we estimated the cost to eat vegetables that were bought fresh in one or both ways—raw (fig. 6b) and/or cooked (fig. 6c).

After adjusting for inedible parts and accounting for cooking yields, if applicable, 11 types of fresh vegetable could be consumed (raw and/or cooked) for less than 50 cents per cup equivalent. The least expensive to consume were boiled potatoes (19 cents per cup equivalent), raw whole carrots (25 cents), iceberg lettuce (26 cents), boiled cabbage (27 cents), and raw onions (28 cents). The most expensive was boiled leaf and baby spinach at $2.02 per cup equivalent.

Households may be able to buy some random-weight vegetables for less money than the higher value-added, fresh-cut produce we priced. For example, bunches of spinach may cost less than the bagged leaf and baby spinach priced for this study. The same may be true for Romaine lettuce, mustard greens, collard greens, turnip greens, kale, and winter squash. As discussed earlier, data limitations prevented us from pricing these foods as sold on a random-weight basis.

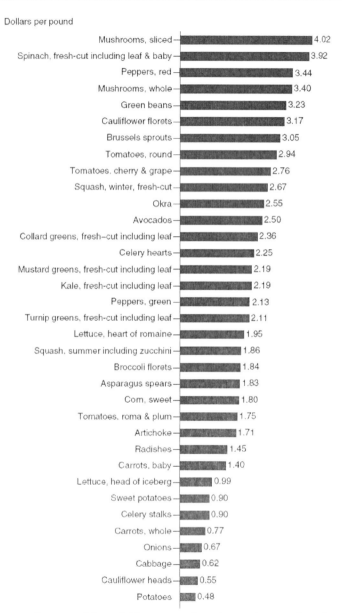

Note: Prices are for vegetables sold in a prepackaged container, such as in a bag or clamshell, and vegetables sold on a count basis, such as iceberg lettuce and cauliflower priced per head.

Source: USDA, Economic Research Service analysis of 2008 Nielsen Homescan data.

Figure 6a. Fresh vegetables: Average retail prices.

How Much Do Fruits and Vegetables Cost? 127

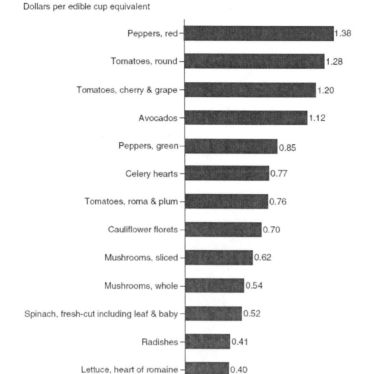

Note: Prices are for vegetables sold in a prepackaged container, such as in a bag or clamshell, and vegetables sold on a count basis, such as iceberg lettuce priced per head. Edible cup equivalents are the units of measurement for fruit/vegetable-consumption recommendations.

Source: USDA, Economic Research Service analysis of 2008 Nielsen Homescan data.

Figure 6b. Fresh vegetables: Average prices per edible cup equivalent(consumed raw).

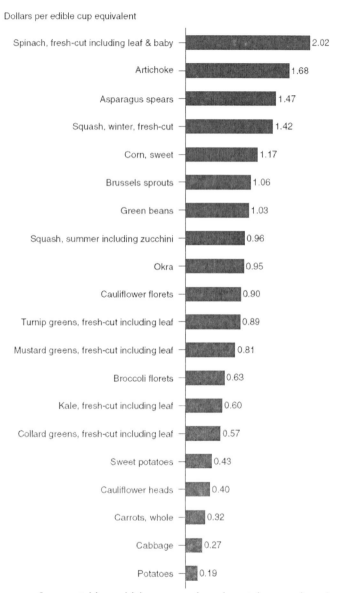

Note: Prices are for vegetables sold in a prepackaged container, such as in a bag or clamshell, and vegetables sold on a count basis, such as cauliflower priced per head. Edible cup equivalents are the units of measurement for fruit/vegetable-consumption recommendations.
Source: USDA, Economic Research Service analysis of 2008 Nielsen Homescan data.
Figure 6c. Fresh vegetables: Average prices per edible cup equivalent(cooked).

Canned Vegetables

The sample of canned vegetables in our study included 24 products. Prices ranged from 67 cents per pound for cut green beans to $3.12 per pound for olives (fig. 7a). Some relatively inexpensive canned vegetables included whole kernel corn, sliced carrots, potatoes, and green peas.

Households are assumed to drain and discard the liquid in which canned vegetables are packed. The vegetables may thereafter be cooked or simply eaten without further preparation. For this study, we use the weight for a cup equivalent reported in the MPED for canned vegetables that are eaten without further preparation. After adjusting for drainage and converting to a cup equivalent basis, 10 of the 24 canned vegetable products cost less than 50 cents to eat (fig. 7b). Sauerkraut was the least expensive canned vegetable to eat at 30 cents per cup equivalent, followed by cut green beans (34 cents), sliced carrots (34 cents), and whole kernel, sweet corn (37 cents).

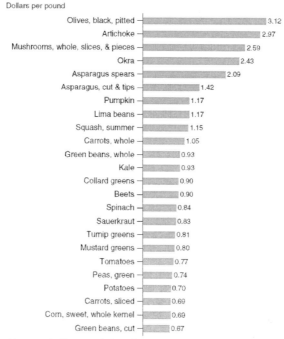

Source: USDA, Economic Research Service analysis of 2008 Nielsen Homescan data.
Figure 7a. Canned vegetables: Average retail prices.

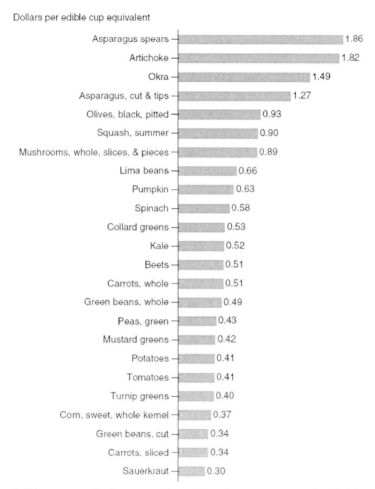

Note: Edible cup equivalents are the units of measurement for fruit/vegetable-consumption recommendations.
Source: USDA, Economic Research Service analysis of 2008 Nielsen Homescan data.
Figure 7b. Canned vegetables: Average prices per edible cup equivalent.

Frozen Vegetables

Our study included 23 frozen vegetables which ranged in price from 93 cents per pound for french fries to $5.11 per pound for artichokes (fig. 8a). Some relatively inexpensive frozen vegetables to buy included carrots, cut

How Much Do Fruits and Vegetables Cost? 131

green beans, okra, green peas, collard greens, whole kernel corn, and cauliflower.

We assume that most frozen vegetables, such as whole kernel corn, are boiled prior to consumption though some are thawed or baked. French fries, for example, are baked. Under these assumptions, four types of frozen vegetable cost less than 50 cents per cup equivalent. Cut green beans were the least expensive frozen vegetable to consume at 37 cents per cup equivalent, followed by frozen carrots (39 cents), french fries (41 cents), and kale (48 cents) (fig. 8b).

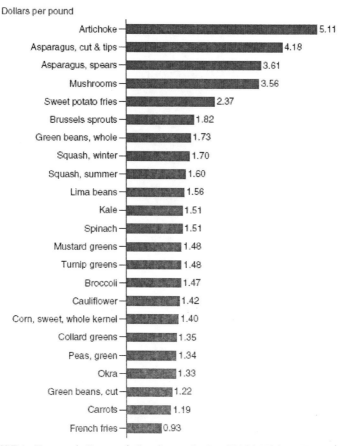

Source: USDA, Economic Research Service analysis of 2008 Nielsen Homescan data.
Figure 8a. Frozen vegetables: Average retail prices.

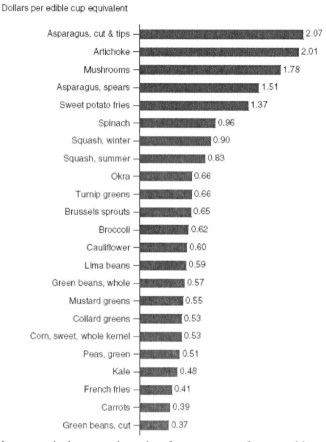

Note: Edible cup equivalents are the units of measurement for vegetable-consumption recommendations.
Source: USDA, Economic Research Service analysis of 2008 Nielsen Homescan data.
Figure 8b. Frozen vegetables: Average prices per edible cup equivalent.

Beans and Peas

We priced 12 canned and dried beans as well as dried lentils. Beans and peas tend to cost less per pound to buy, but are relatively more expensive per cup equivalent to eat. However, on an edible basis, all 12 products cost less than 50 cents per cup equivalent to consume. Of the beans and peas we priced, dried pinto beans were the least expensive to eat at 13 cents per cup equivalent followed by lentils at 15 cents per cup equivalent (figs. 9a and 9b).

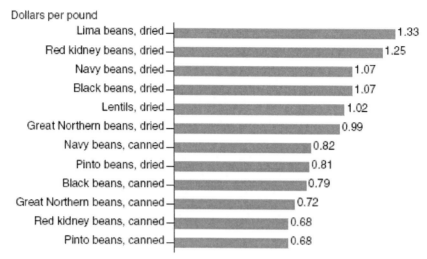

Source: USDA, Economic Research Service analysis of 2008 Nielsen Homescan data.
Figure 9a. Beans and peas: Average retail prices.

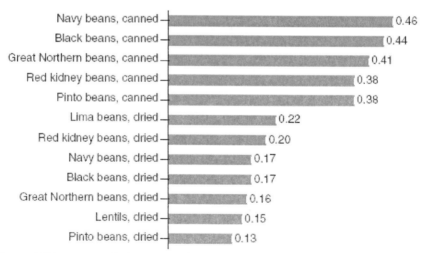

Note: Edible cup equivalents are the units of measurement for fruit/vegetable-consumption recommendations.
Source: USDA, Economic Research Service analysis of 2008 Nielsen Homescan data.
Figure 9b. Beans and peas: Average prices per edible cup equivalent.

DISCUSSION

Having estimated average retail prices and prices per edible cup equivalent for 153 fresh and processed fruits and vegetables, we find much variation from the least expensive to most expensive products. As our results in figures 1-9 demonstrate, an edible cup equivalent of fruit can cost 17 cents for fresh watermelon or $2.06 for fresh raspberries. Similarly, an edible cup equivalent of vegetables can cost 13 cents for dry pinto beans or $2.07 for frozen asparagus cuts and tips.

We further find that neither fresh nor processed foods are a consistently cheaper way to eat fruits and vegetables. For example, fresh whole carrots eaten raw (25 cents per cup equivalent) are less expensive to consume than either canned carrots (34 cents per cup equivalent) or frozen carrots (39 cents per cup equivalent). By contrast, canned peaches (58 cents per cup equivalent) are more economical than fresh (66 cents per cup equivalent).

Retail prices also appear to be a poor indicator of prices per edible cup equivalent. Fresh broccoli florets and fresh ears of sweet corn both sell for around $1.80 per pound at retail stores, on average. However, after boiling and removing inedible parts, sweet corn ($1.17 per cup equivalent) costs almost twice as much as broccoli florets (63 cents per cup equivalent). Similarly, fresh apples and fresh pineapple both sell for slightly more than $1 per pound at retail stores. However, on a cup-equivalent basis, the apples are much cheaper (28 cents versus 70 cents).

Tables 5 and 6 show some of the less expensive products by subgroup. Many types of fruit juice and a few types of whole fruit cost less than 30 cents per cup equivalent including fresh watermelon, fresh bananas, and fresh apples. Several vegetables in different subgroups were also available in this price range. Whole carrots (25 cents per cup equivalent) were the cheapest red or orange vegetable. Potatoes were the cheapest starchy vegetable (19 cents per cup equivalent). Many types of beans and peas cost less than 20 cents per cup equivalent. Dark green vegetables were slightly more expensive. Fresh Romaine hearts and canned turnip greens were the least expensive dark green vegetables to consume (40 cents per cup equivalent).

To illustrate how much it costs to satisfy overall fruit and vegetable guidelines, we provide a 3-day example for a person on a 2,000-calorie daily diet (table 7). Each daily example includes the recommended 4.5 cup equivalents for this person (2 cup equivalents of fruit and 2.5 cup equivalents of vegetables). We also include popular foods from different fruit and vegetable subgroups. Although total costs for consuming the foods in the examples vary

How Much Do Fruits and Vegetables Cost? 135

from day to day, in 2008, it was possible to satisfy recommendations in the *2010 Dietary Guidelines for Americans* for about $2 to $2.50 per day, or approximately 50 cents per edible cup equivalent.

Table 5. Less expensive fruit, 2008

Fruit type	Average price (*cents/edible cup equivalent*)
Whole and cut fruit	
Apples—fresh	28
Applesauce—canned	46
Bananas—fresh	21
Grapes—dried (raisins)	39
Honeydew melon—fresh	45
Nectarines—fresh	49
Oranges, navel—fresh	34
Pears—fresh	42
Pineapple—canned	49
Plums—fresh	48
Watermelon—fresh	17
Juice	
Apple	20
Grape	25
Grapefruit	27
Orange	26

Note: All juice prices shown above are for products sold at retail stores in the form of frozen concentrate and then reconstituted by the consumer at home.

Source: USDA, Economic Research Service analysis of 2008 Nielsen Homescan data.

Table 6. Less expensive vegetables, 2008

Vegetable type	Average price (cents/edible cup equivalent)
Dark green	
Kale—frozen	*48*
Mustard greens—canned	42
Mustard greens—frozen	55
Romaine lettuce—fresh	40
Turnip greens—canned	40
Red and orange	
Carrots—whole, fresh	25
Carrots—baby, fresh	40
Vegetable type	Average price (cents/edible cup equivalent)
Carrots—canned, cut	34
Carrots—frozen	39
Sweet potatoes—boiled from fresh	43
Tomatoes—canned, whole and cut	41
Beans and peas	
Pinto beans—boiled	13
Great Northern beans—boiled	16
Navy beans—boiled	17
Black beans—boiled	17
Red kidney beans—boiled	20
Starchy	
Corn—canned, whole kernel	37
Potatoes—boiled from fresh	19
Green peas—canned	43
Other	
Cabbage—boiled from fresh	27
Cauliflower—boiled from fresh	49
Celery stalks—fresh	33
Green beans—canned, cut and sliced	34
Green beans—frozen, not whole	37
Iceberg lettuce—fresh heads	26
Onions—fresh	28
Radish—fresh	41
Sauerkraut—canned	30

Note: All bean prices are for products bought in dried form and then boiled by the consumer at home.

Source: USDA, Economic Research Service analysis of 2008 Nielsen Homescan data.

Table 7. Costs for meeting vegetable and fruit recommendations in the *2010 Dietary Guidelines for Americans*, daily examples[1]

Fruit or vegetable type	Cup	Day 1 Cost (cents)	Cup	Day 2 Cost (cents)	Cup	Day 3 Cost (cents)
Fruit and fruit juice						
Orange juice, from frozen concentrate	1	26	1	26	1	26
Strawberries, fresh	1	89				
Cantaloupe, fresh			0.5	32		
Banana, fresh			0.5	11		
Raisins					0.5	20
Apples, fresh					0.5	14
Dark green vegetables						
Romaine lettuce, fresh	0.5	20				
Spinach, cooked from frozen			0.25	24		
Broccoli florets, cooked from fresh					0.5	32
Red and orange vegetables						
Carrots, whole, fresh	0.5	13				
Tomato, Roma, fresh	0.5	38				
Sweet potato, cooked from fresh			0.5	22		
Carrots, cooked from frozen			0.25	10		
Tomato, grape, fresh					0.5	60
Starchy vegetables						
Corn, whole kernel, canned	0.5	19				
Green peas, cooked from frozen			0.5	26		
Potato, cooked from fresh					0.5	10
Other vegetables Onions, fresh	0.25	7				
Green beans, whole, cooked from frozen			0.5	29		
Cauliflower florets, fresh					0.5	35
Celery stalk, fresh					0.5	17
Beans and peas Pinto beans, canned	0.25	10				
Black beans, canned			0.5	22		
Total	4.5	$2.22	4.5	$2.02	4.5	$2.12

[1] For a person on a 2,000-calorie daily diet.

Note: Cups are 1-cup equivalents. Costs are for the number of cup equivalents consumed. Food costs are based on figures 1a-9b. We do not consider how the foods listed here contribute to the intake of micronutrients like potassium, vitamin A, and vitamin C.

Source: USDA, Economic Research Service analysis of 2008 Nielsen Homescan data and Federal dietary recommendations.

REFERENCES

Carlson A., M. Lino, W. Juan, K. Marcoe, L. Bente, H. Hiza, P. Guenther, and E. Leibtag. 2008. *Development of the CNPP Prices Database*, CNPP-22, USDA Center for Nutrition Policy and Promotion. Available at: http://www.cnpp.usda.gov/publications/foodplans/miscpubs/pricesdatabas ereport.pdf/.

Einav, L., E. Leibtag, and A. Nevo. 2008. *On the Accuracy of Nielsen Homescan Data*, ERR-69, USDA, Economic Research Service. Available at: http://www.ers.usda.gov/publications/err69/err69.pdf/.

Guenther P., K. Dodd, J. Reedy, and S. Krebs-Smith. 2006. "Most Americans Eat Much Less Than Recommended Amounts of Fruits and Vegetables," *Journal of the American Dietetic Association*, Vol. 106: pp. 1371-1379.

Reed J., E. Frazão, and R. Itskowitz. 2004. *How Much Do Americans Pay for Fruits and Vegetables?* AIB-790, USDA, Economic Research Service. Available at: http://www.ers.usda.gov/publications/aib790/.

U.S. Department of Agriculture and U.S. Department of Health and Human Services, *Dietary Guidelines for Americans, 2010* (8th ed.). Available at: www.DietaryGuidelines.gov.

U.S. Department of Agriculture, Agricultural Research Service. 1975. *Food Yields Summarized by Different Stages of Preparation*, USDA Agriculture Handbook No. 102. Available at:http://www.ars.usda.gov/sp2userfiles/ place/12354500/data/classics/ah102.pdf/.

U.S. Department of Agriculture, Agricultural Research Service, *National Nutrient Database for Standard Reference* (Release 21). Available at: http://www.ars.usda.gov/.

U.S. Department of Agriculture, Agricultural Research Service, *MyPyramid Equivalents Database*. Available at: http://www.ars.usda.gov/servicesdocs.htm?docid=17558/.

U.S. Department of Agriculture, Center for Nutrition Policy and Promotion, *CNPP Food Prices Database*, 2003-04. Available at http://www.cnpp.usda.gov/usdafoodplanscostoffood.htm/.

U.S. Department of Agriculture, Center for Nutrition Policy and Promotion, MyPyramid.gov. Available at http://www.MyPyramid.gov/.

ACKNOWLEDGMENTS

For their thoughtful comments and assistance, the authors would like to thank Elise Golan, Abebayehu Tegene, Gary Lucier, Fred Kuchler, and Mark Denbaly (Economic Research Service, USDA), Mark Lino, Kristin Koegel, Robert Post, and Patricia Britten (Center for Nutrition Policy and Promotion, USDA), Mykel Taylor (Washington State University), and Diana Cassady (University of California, Davis). Thanks also to Priscilla Smith and Linda Hatcher for editorial assistance and to Cynthia A. Ray for graphic design.

End Notes

[1] Each vegetable subgroup is defined in the *2010 Dietary Guidelines for Americans*. For example, beans and peas are the mature form of legumes, which include kidney beans, pinto beans, black beans, garbanzo beans, lima beans, black-eyed peas, split peas, and lentils. Legumes do not include green beans or green peas.

[2] It is reasonable to question the credibility of Homescan data because households self-report their purchases. Households may make mistakes when reporting information to Nielsen (e.g., some may fail to report all purchases because the recording process is time-consuming). However, validation studies confirm the suitability of these data for calculating average prices paid. Einav et al. (2008) concluded that errors commonly found in Homes-can data should not seriously affect estimates of average prices paid by all households. They also found that errors in the Homescan data are of the same order of magnitude as reporting errors in data sets commonly used to measure earnings and employment, for example.

[3] USDA's nutrition education guidance system, known as MyPyramid (MyPyramid.gov), helps nutrition educators, health professionals, and consumers implement the Dietary Guidelines for Americans (DGA) through personalized eating plans and interactive tools to help them plan/ assess the best food choices based on the DGA. The DGA, through MyPyramid, recommends amounts of food groups, including vegetables and fruits, to meet the nutrient needs of Americans ages 2 and older, at various calorie levels, to help reduce the risk of chronic illness and maintain a healthy weight

INDEX

A

accounting, 5, 7, 8, 9, 13, 22, 24, 25, 29, 40, 112, 125
adjustment, 69, 70, 81, 86, 87, 88
age, 109
agencies, 64
aggregation, 14, 23, 54
agribusiness, 3, 7, 24, 25, 26, 27, 28, 29, 49, 65
algorithm, 60, 65
apples, 110, 111, 117, 123, 134
asparagus, 106, 134
assets, 5, 7, 21, 26, 27, 28, 59, 65
asymmetry, 81
average costs, 6, 25, 116
awareness, viii, 105, 106, 108

B

BEA, 5, 14, 15, 22, 27, 54, 55, 59, 60, 65, 66
beef, 21, 67, 68, 69, 70, 71, 72, 75, 76, 77, 78, 80, 82, 85, 86, 87, 88, 89, 90, 91, 92, 97, 98, 99, 102, 103
behaviors, 81
benefits, 9
beverages, 5, 14, 15, 19, 20

bias, 54
bounds, 72, 74, 81, 82, 89, 97, 103
building blocks, 55
Bureau of Labor Statistics, 5, 14, 22, 48, 53, 67, 68, 71, 73, 74, 76, 77, 78, 83, 84, 85, 90, 91, 98, 99, 100, 101, 102
buyers, 9

C

cabbage, 125
calorie, 105, 106, 108, 134, 137, 139
cane sugar, 15
capital markets, 65
categorization, 82
cation, 54
cattle, 9, 12, 68, 69, 71, 79, 86, 87, 92, 99, 102
Census, 7, 14, 48, 53, 54, 56, 65
chemical, 61
chemical industry, 61
chronic diseases, 108
chronic illness, 139
clarity, 10
cleanup, 17
clients, 23
collusion, 49
commodity, 2, 5, 8, 9, 10, 11, 12, 13, 14, 17, 18, 21, 22, 29, 41, 50, 54, 55, 56, 57, 58,

59, 60, 61, 64, 65, 68, 69, 70, 71, 72, 75, 80, 81, 86, 92
compensation, 9, 27
compilation, 54
complexity, 70, 80
composition, 28
computation, 28, 55
concordance, 54
Congress, 6, 13
consolidation, 49, 79
consumer price index, 65
Consumer Price Index, 70, 72, 74, 84, 93, 102
consumers, vii, 2, 3, 6, 18, 21, 48, 70, 75, 108, 111, 122, 123, 139
consumption, viii, 14, 61, 78, 105, 106, 107, 108, 110, 111, 113, 117, 119, 120, 121, 122, 123, 124, 127, 128, 130, 131, 132, 133
convention, 12, 13, 54
convergence, 86
cooking, 107, 111, 115, 116, 117, 123, 125
cost, viii, 5, 6, 8, 18, 24, 27, 48, 62, 64, 65, 67, 68, 72, 106, 107, 108, 109, 111, 112, 116, 117, 118, 121, 123, 125, 129, 131, 132, 134
covering, 73
CPI, 65, 71, 72, 73, 74, 84, 98, 100, 102
critical value, 96
crops, 15, 102
crude oil, 41
current account, 11

D

data processing, 53
data set, 139
decomposition, 62
deduction, 23, 50, 65
Denmark, 52
Department of Agriculture, 1, 2, 50, 52, 53, 67, 105, 108, 138
Department of Health and Human Services, 108, 138

dependent variable, 97, 98, 99, 100, 101, 102
depth, 77
deviation, 74, 88, 91
diet, viii, 15, 105, 106, 108, 120, 134, 137
Dietary Guidelines for Americans, viii, 105, 106, 108, 109, 135, 137, 138, 139
dietary habits, 108
direct measure, 23, 53
distribution, 1, 3, 72
divergence, 21, 86
domestic industry, 5, 7, 11, 61
domestic labor, 27
double counting, 12, 62
drainage, 129

E

earnings, 27, 139
ECM, 79, 80, 82, 85, 86, 87, 88, 90, 91, 97, 98, 99, 100, 101, 103
Economic Research Service (ERS), vii, 2, 6
economies of scale, 41, 49
education, 108, 139
educators, 139
electricity, 25, 54, 56, 57, 88, 101
empirical studies, 70, 80
employees, 14, 65
employers, 14, 65
employment, 65, 139
energy, 3, 7, 25, 26, 29, 40, 41, 49, 54, 65, 70, 71, 79, 88
energy input, 71
energy prices, 49
entropy, 60
environment, 73
equilibrium, 69, 80
equilibrium price, 69
equipment, 9, 11, 65
estimation process, 11
evidence, 80
exclusion, 10
exercise, 109

Index 143

expenditures, 1, 5, 8, 9, 10, 11, 12, 14, 16, 17, 18, 21, 22, 23, 28, 29, 40, 49, 50, 55, 58, 61, 64, 65, 115

F

farm share, 2, 3, 5, 6, 8, 9, 10, 11, 13, 15, 16, 17, 18, 19, 20, 21, 22, 26, 49, 50, 53, 72, 75, 77
farmers, vii, 2, 6, 10, 14
farms, 10
Federal dietary guidance, viii, 105, 106, 107
Federal Register, 64
fertilizers, 11, 65
field crops, vii, 68
financial, 3
flexibility, 70, 71
flour, 10, 11, 69, 71, 72, 73, 74, 80, 82, 85, 86, 87, 88, 89, 92, 97
fluctuations, 6, 88
food dollar, vii, 1, 2, 3, 5, 6, 7, 8, 9, 10, 11, 12, 13, 15, 16, 17, 19, 20, 21, 22, 23, 24, 26, 27, 28, 29, 30, 40, 41, 42, 49, 50, 53, 54, 58, 59, 61, 65, 75
food industry, 1, 3, 50
food processing industry, 40
food production, 6, 24
food products, 10, 21, 23, 49, 61, 71, 75, 86, 87, 110, 112
food services, 5, 7, 18, 41, 48, 49
forecasting, 93
formation, 25
freedom, 70
fruits, viii, 10, 105, 106, 107, 108, 109, 110, 112, 115, 116, 117, 121, 122, 123, 134, 139
funds, 28

G

GDP, 65
GNP, 94
goods and services, 8, 14, 17
grouping, 81, 97

growth, 18, 48, 79
guidance, viii, 64, 105, 106, 107, 139
guidelines, 134

H

health, viii, 105, 106, 108, 139

I

identity, 12
import prices, 65
imports, 1, 3, 5, 7, 10, 11, 27, 29, 40, 41, 49, 61, 65
impulses, 103
income, viii, 5, 9, 27, 28, 29, 65, 105, 106, 108
income tax, 29, 65
indexing, 14
individuals, 1
industries, 9, 11, 14, 22, 23, 24, 25, 26, 27, 28, 29, 40, 54, 55, 56, 57, 62, 63, 64, 65, 71, 77, 78
industry, 1, 2, 3, 5, 7, 8, 9, 11, 12, 13, 14, 21, 22, 23, 24, 26, 27, 28, 29, 40, 49, 53, 54, 55, 56, 57, 58, 59, 61, 62, 63, 64, 65, 66, 78, 81
industry value added, 2, 27, 54
inertia, 81, 84
inflation, 93
information processing, 65
ingredients, 3, 8, 10, 11, 29, 49, 65
input-output analysis, 2, 8, 11, 55
institutions, 14, 65
integration, 95
intellectual property, 9
inversion, 63, 65
issues, 7, 81

J

JO-based food dollar series, 24

144   Index

K

kidney, 114, 136, 139

L

labor market, 49
labor markets, 49
lead, 9, 10, 81, 82, 88, 89
light, 1, 80
livestock, 77, 80
locus, 79, 102
lower prices, 116

M

machinery, 9, 65
magnitude, 69, 80, 82, 84, 85, 86, 87, 88, 89, 139
major field crops, vii, 68
majority, 108
Mandarin, 113, 117, 118
manufacturing, 10, 27
mapping, 55
market segment, 6, 8
market structure, vii, 6, 81
marketing, vii, 1, 2, 3, 5, 6, 7, 8, 9, 10, 14, 15, 16, 17, 18, 21, 24, 27, 50, 67, 68, 71, 81, 84, 92
marketing bill, vii, 1, 2, 3, 5, 6, 7, 8, 9, 10, 14, 15, 17, 18, 21, 24, 27, 50
Maryland, 64
materials, 23, 50
mathematical programming, 60
matrix, 11, 12, 13, 14, 22, 23, 28, 53, 54, 55, 57, 58, 61, 62, 63, 65, 66
matrix algebra, 12
measurement, 1, 8, 27, 53, 119, 120, 123, 124, 127, 128, 130, 132, 133
meat, 78, 102
melon, 110, 135
methodology, 15, 49, 80, 109
micronutrients, 137
model specification, 92

models, 68, 70, 71, 82, 85, 86, 87, 88, 89, 90, 92, 93, 96, 97, 103
multiplication, 12

N

natural gas, 25, 54, 56, 57, 65
natural resources, 9, 65
neutral, 64
nitrogen, 65
nonlinear impulse response function (NLIRF), 88
North America, 54
North American Industry Classifi cation System (NAICS), 54
null, 55, 95, 97
null hypothesis, 95, 97
nutrient, 139
nutrition, 108, 139

O

Office of Management and Budget, 64
oil, 112
operations, 9, 79, 102
organize, 62
output method, 2
overlap, 15

P

participants, 21
partition, 62
percentile, 72
petroleum, 11, 41
physical activity, 109
physical structure, 9, 65
policy, 64
population, 6
population growth, 6
potassium, 137
potato, 137
poultry, 9
preparation, 17, 18, 75, 116, 117, 129

price changes, 67, 68, 69, 70, 71, 72, 80, 82, 84, 85, 87, 88, 89, 92, 103
price index, 18, 21, 65, 86
pricing behavior, 92
primary data, 54
primary factor value added, 2, 57, 64
Producer Price Index, 71, 72, 74, 98, 99, 100, 101
producers, 6, 9
production costs, 29
professionals, 139
property taxes, 29

Q

quality standards, 64

R

raw farm commodities, vii, 2
reasoning, 25
recalling, 22
recommendations, viii, 105, 106, 108, 109, 113, 119, 120, 122, 123, 124, 127, 128, 130, 132, 133, 135, 137
regression, 86, 89, 97, 98, 99, 100, 101, 103
regression model, 97, 98, 99, 100, 101
relative prices, 18
repetitions, 97
requirements, 63
residuals, 95
resources, 9
response, 67, 68, 69, 70, 71, 72, 73, 75, 80, 81, 82, 85, 86, 87, 88, 90, 91, 92, 96, 102, 103
response time, 81
responsiveness, 68, 87
restaurants, 8, 14, 21, 24, 41, 48, 49
restrictions, 97
retail, vii, 7, 40, 41, 49, 50, 54, 56, 57, 58, 65, 66, 67, 68, 69, 70, 71, 72, 73, 75, 76, 77, 78, 80, 81, 82, 83, 84, 85, 86, 87, 88, 89, 90, 91, 92, 93, 97, 98, 100, 102, 103, 105, 106, 107, 108, 109, 110, 111, 112, 115, 116, 117, 118, 120, 121, 122, 124, 125, 126, 129, 131, 133, 134, 135
risk, 108, 139
root, 95
roots, 112

S

scale economies, 49
scaling, 66
school, 14
scope, 1
Secretary of Agriculture, 6
service industries, 56
services, 3, 6, 7, 9, 11, 17, 18, 21, 23, 24, 26, 40, 41, 48, 49, 50, 54, 56, 64, 138
shape, viii, 105, 106, 108
showing, 72, 85, 92
simulation, 88
simulations, 97
skin, 116
solution, 60
spending, vii, 2
stability, 95
standard deviation, 72, 90
state, 23, 122, 123
states, 53, 63
statistics, 7, 65
storage, 10, 81
structural changes, 96
structure, vii, 6, 8, 23, 28, 53
subgroups, 108, 134
supply chain, vii, 1, 2, 3, 5, 8, 9, 10, 21, 22, 23, 24, 25, 26, 27, 28, 29, 40, 50, 62, 63, 64, 67, 68, 70, 71, 72, 75, 78, 80, 81, 84, 86, 87, 89, 92, 95
supply chain analysis, 2, 8, 21, 24
surging, 87

T

target, 55
taxes, 1, 3, 9, 27, 28, 29, 40, 59, 65
techniques, 96

technologies, 54, 65
technology, 12
test statistic, 96
testing, 96, 97
time periods, 73, 74
time series, 53, 95, 96
Title V, 6
total costs, 64, 134
total energy, 8
trade, 7, 23, 41, 54, 56, 57, 58, 59, 65, 66
transaction costs, 81
transactions, 13, 15, 50, 66
transformation, 81
transmission, 68, 71, 80
transport, 10, 59, 71
transport costs, 71
transportation, 3, 7, 11, 25, 26, 27, 57, 58

U

U.S. Department of Commerce, 48, 53, 54
U.S. Department of Labor, 22, 48, 53, 73,
74, 76, 77, 78, 83, 84, 85, 90, 91, 98, 99,
100, 101
uniform, 71, 85
unit cost, 18
United States, 1, 9, 10, 41, 67, 78, 105, 108,
115
USDA, vii, 2, 6, 7, 16, 19, 20, 26, 39, 47,
48, 49, 50, 51, 53, 64, 72, 73, 74, 76, 77,
78, 79, 83, 84, 85, 90, 91, 93, 94, 98, 99,
100, 101, 102, 107, 108, 111, 113, 114,
115, 116, 118, 119, 120, 121, 122, 123,
124, 126, 127, 128, 129, 130, 131, 132,
133, 135, 136, 137, 138, 139
utility costs, 8

V

validation, 139
value-added requirement, 64
value-added requirements, 64
variables, 70, 71, 72, 73, 75, 76, 78, 79, 87,
95, 102, 103
variations, 70, 73
varieties, 10, 111, 120, 121
vector, 11, 12, 22, 28, 55, 56, 57, 58, 61, 63,
64, 65
vegetables, viii, 105, 106, 107, 108, 109,
110, 112, 113, 114, 115, 116, 125, 126,
127, 128, 129, 130, 131, 132, 134, 136,
137, 139
vitamin A, 137
vitamin C, 137
volatility, 41, 70, 72, 74, 82

W

wages, 9, 28
Washington, 139
water, 61, 65, 122, 123
web, 102
wholesale, vii, 23, 54, 56, 57, 58, 65, 66, 67,
68, 69, 70, 71, 72, 73, 76, 77, 78, 82, 83,
84, 85, 86, 87, 88, 89, 90, 91, 92, 97, 98,
99, 100, 101, 102, 103, 107, 109
wholesale-to-retail price, 67, 68, 69, 76, 77,
84, 86
workers, 27, 28